THE THEORY AND PRACTICE OF WORM GEAR DRIVES

ILLÉS DUDÁS

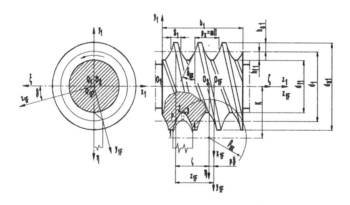

Series Consultant: Prof KJ Stout, University of Huddersfield, UK

THE THEORY AND PRACTICE OF WORM GEAR DRIVES

ILLÉS DUDÁS

Department of Production Engineering, University of Miskolc, Hungary

PENTON PRESS, LONDON

Penton Press
Kogan Page Ltd
120 Pentonville Road
London N1 9JN
www.kogan-page.co.uk

© Illés Dudás 2000

British Library Cataloguing in Publication Data

A CIP record for this book is available from the British Library

ISBN 9781 9039 9661 4

Typeset by The Midlands Book Typesetting Company Ltd, Loughborough, Leicestershire, England.

Printed and bound by CPI Antony Rowe, Eastbourne

To my wife, three children and parents

The author with his early CNC grinding wheel dressing experimental equipment

CONTENTS

Foreword *by Professor F.L. Litvin, University of Illinois*   **xi**

Preface   **xiii**

Acknowledgements   **xvi**

List of symbols   **xix**

1	**Introduction**	**1**
	1.1 Classification of worm gear drives	4
2	**A short history and review of the literature**	**7**
	2.1 A short history of the worm gear drive	7
	2.2 Development of tooth cutting theory for spatial drives	13
	2.3 Cylindrical worm surfaces	16
	2.3.1 Helicoidal surfaces having arched profile	16
	2.3.2 Cylindrical worm gear drives with ruled surfaces	26
	2.4 Conical helicoid surfaces	26
	2.5 Surface of tools	29
	2.6 General conclusions based on the literature	30
3	**Manufacturing geometry for constant pitch helicoidal surfaces**	**33**
	3.1 Development of manufacturing of cylindrical worm gear drives having arched profile	33
	3.1.1 Analysis and equation of helicoidal surface having circular profile in axial section	35
	3.1.2 Analysis of worm manufacturing finishing; an exact solution	42

3.1.3 Problems of manufacturing geometry
during final machining of worm –
determination of grinding wheel profile   44

3.2   Investigation of geometric problems in
manufacturing cylindrical helicoidal surfaces
having constant lead; general mathematical –
kinematic model   61

3.2.1 Investigation of geometric problems when
manufacturing cylindrical helicoid surfaces
using general mathematical – kinematic
model   64

3.2.2 Analysis of manufacturing geometry for
conical helicoid surfaces   75

3.3   Geometric analysis of hobs for manufacturing
worm gears and face-gears mated cylindrical
or conical worms   102

3.3.1 Investigation of cutting tool for
manufacturing worm gear mated with
worm having arched profile   110

4.   **General mathematical model for investigation of hobs
suitable for generating cylindrical and conical worms,
worm gears and face gear generators**   **124**

4.1   Application of general mathematical – kinematic
model to determine surface of helicoidal surface-
generating tool for cylindrical thread surfaces   135

4.2   Machining geometry of cylindrical worm gear
drive having circular profile in axial section   136

4.3   Machining geometry of spiroid drives   148

4.4   Intersection of cylindrical helicoidal surface
having circular profile in axial section (ZTA)
and the Archimedian thread face surface
as generating curve of back surface   162

4.4.1 Generation of radial back surface with
generator curve   164

4.4.2 Contact curve of the back surface and the
grinding wheel   165

4.5   Manufactured tools for worm gear generation
and other tools having helicoidal surfaces   169

4.5.1 Design and manufacture of worm gear
milling cutters   169

5	**Grinding wheel profiling devices**	**182**
	5.1 Devices operated according to mechanical principle	183
	5.2 Advanced version of the wheel-regulating device operating on the mechanical principle	186
	5.3 CNC-controlled grinding wheel profiling equipment for general use	191
6	**Quality control of worms**	**200**
	6.1 Checking the geometry of worms	200
	6.1.1 Determination of worm profile deviation	201
	6.2 Checking of helicoidal surfaces on 3D measuring machines	204
	6.2.1 Use of 3D measuring machines	206
	6.3 Checking of helicoidal surfaces by application of 3D measuring device prepared for general use (without circular table, CNC-controlled)	209
	6.4 Results of measurement of helicoidal surfaces	217
7	**Manufacture of helicoidal surfaces in modern intelligent integrated systems**	**222**
	7.1 Application of expert systems to the manufacture of helicoidal surfaces	222
	7.1.1 Problems of manufacturing worm gear drives	223
	7.1.2 Structure of the system	224
	7.1.3 The full process	224
	7.2 Intelligent automation for design and manufacture of worm gear drives	227
	7.2.1 Conceptual design of helicoidal driving mates	228
	7.2.2 Manufacture of worms and worm gears	245
	7.3 Measurement and checking of helicoidal surfaces in an intelligent system	251
	7.3.1 Checking of geometry using coordinate measuring machine	253
	7.4 Development of the universal thread-grinding machine	255
	7.4.1 Review of thread surfaces from the point of view of thread-grinding machines	255
	7.4.2 Manufacturing problems of thread surfaces	255

		7.4.3	Requirements of the thread-grinding machine	257
		7.4.4	Development of a possible version	258
	7.5	Conclusions		259
8.	**Main operating characteristics and quality assessment of worm gear drives**			**260**
	8.1	Testing the meshing of the mated elements		260
		8.1.1	Building in the mating elements	261
		8.1.2	Adjustment and position checking of contact area	262
	8.2	Checking the important operational characteristics of worm gear drives		271
		8.2.1	Running in of the drives	271
		8.2.2	Determination of optimal oil level	274
		8.2.3	Investigation of warming up of the drives	274
		8.2.4	Investigation of efficiency of drives	277
		8.2.5	Investigation of noise level of drives	280
9	**Summary of results of research work**			**289**
References				**294**
Further reading				**303**
Index				**331**

FOREWORD

The writing of this Foreword to this book presents me with a wonderful opportunity to recall my visits to Miskolc and my meetings with the distinguished scientists and the friends that I was lucky enough to make there.

My friends from Miskolc, Professor Zeno Terplan and Dr Jozsef Drobni, gave me the best present that I could have asked for – they translated in 1972 the Russian edition of my book *Theory of Gearing* into Hungarian.

I was delighted to find in my conversations with Drs Imre Levai, Zeno Terplan and Illés Dudás a mutual interest in topics such as non-circular gears, planetary trains and worm gear drives.

The greatest reward for a scientist is to have a following, and this I found in Hungary.

My joy in this could perhaps best be expressed by citing the famous verse 'The Arrow and the Song' by Henry Wadsworth Longfellow:

I shot an arrow into the air,
It fell to earth, I knew not where;

And the song, from beginning to end,
I found again in the heart of a friend

I hope that this short introduction explains why I am grateful for the opportunity to write a Foreword to this excellent book written by Professor Dudás.

The generation and manufacture of worm gear drives and the design of tools (hobs, grinding disks) for worm and worm gear generation is an important area of research. The application of CNC machines to the manufacture of worms and worm gears, their precision testing, and the computerized design of tools have broadened the horizons of research and have required from the researchers a good knowledge of the theory of gearing and specialized topics in differential geometry.

In this book Dr Illés Dudás makes a significant contribution to these topics of research; included are the author's summaries of the results of research obtained by himself and other researchers. In addition, Professor Dudás demonstrates the results of his great and wide experience in the design and manufacture of worm gear drives and in neighbouring subject areas.

The contents of the book cover the main topics of the design and manufacture of gear drives. I am familiar with the research performed by Professor Dudás whom I was able to meet at International Conferences (in San Diego and Dresden) and at our University, and by exchange of our publications.

There is little doubt that this book will be prove to be a most useful work for researchers and engineers in the area of gears.

Faydor L. Litvin
University of Illinois at Chicago
Chicago, USA
1999

PREFACE

Automation is playing an ever-increasing role in the development
of both product and manufacturing technologies. Automation provides important means of improving quality and increasing
productivity as well as making production more flexible, in line with
changing needs. State of art computer control now has a role for
machine tools and in manufacturing technology. Design of the product as well as of manufacturing equipment has been taken over by
computer-aided, and sometimes by completely automated, systems.
In the increase in efficiency of manufacturing processes and product quality, the most important element has been computer-aided
engineering.

Helicoid surfaces are often used in mechanical structures like
worm gear drives, power screws, screw pumps and screw compres-
sors, machine tools, and generating gear teeth. Therefore many
research and manufacturing organizations are becoming involved
with their design, manufacture, quality control and application.

Theory and practice in this field are usually treated separately in
textbooks. There are significant differences between different machining technologies, and checking methods for helicoidal surfaces
are not always designed and manufactured precisely and optimally.

I have been particularly fortunate to have been able to work,
during the course of my career, in many fields of engineering. During my years as a professional engineer I always felt attached to
scientific investigation concerned with the correlation between construction and manufacturing technology. Following a short period
in industrial practice I worked, for ten years, as a designer. My first
assignments were the design of service equipment (for example the
DKLM-450 type wire-rope bunch lifter), and later, wire pulling

stages, wire–end sharpeners, etc. The need for an improved worm gear drive arose in the course of this work.

The machine factory at Diósgyör (DIGÉP, Hungary) was using wire pulling stages and decided to modernize them, to reduce their noise level, weight and cost along with developing an increase in the efficiency and load-carrying capacity. The modernization was carried out successfully so that the kinematically complicated drive systems were simplified too.

The experience gained during tests showed that drive systems fulfilling exacting requirements can only be solved by using special worm gear drives. The technical development of worm gear drives at DIGÉP resulted in worm gear drives with different geometries such as convolute helicoids with limited bearing capacity, worm drives with rolling contact elements and helicoidal surfaces curved at their axial section. Comparing them, it became clear that the development of curved axial section type helicoidal surfaces was called for.

Research in the fields of manufacturing technology development, as well as toothing geometry of mated pairs and the overall checking and quality control of these drives, are summarized in some of my published works (Dudás, 1973, 1980, 1988b).

Worm gear drives designed and manufactured by application of this newly developed method have operated efficiently both in Hungary and abroad in a range of different products.

In my present position as Head of Department of Production Engineering at the University of Miskolc, it has been possible to continue my previous research work in this field, to fill gaps in the work and to search for a possible description of their generalized geometry, starting from the common characteristics of the different types of helicoidal surfaces.

This book basically aims to clear up geometrical problems arising during manufacture and provide theoretical equations necessary to solve them, thus filling a gap existing in publications in the field.

In the nine chapters of the book, both theory and practice are covered. The contents may be summarized as follows:

1. This introductory chapter provides the reader with a view of the aim of the book and provides a short review of the history of worm gear drives.
2. An analysis of the literature of the subject and a summary of conclusions to be drawn from it concerning the field covered by the book.

3. The theory needed for the proper geometrical manufacture of helicoidal surfaces of constant pitch is introduced here. Further, a mathematical model is developed, suitable for the handling of the production geometry of both cylindrical and conical helicoidal surfaces.

4. Introduces a general mathematical kinematic model for both cylindrical and conical worm gear drives, as well as the design and manufacture of the necessary tools.

5. Newly patented truing equipment for grinding wheels is described.

6. Checking methods for helicoidal surfaces are described.

7. Design methods and the procedures for manufacturing helicoidal surfaces using intelligent CIM systems are introduced.

8. The basic theory, operational characteristics and the possible applications for drives are summarized in this chapter.

9. Comprises a summary of results of research work.

10. A full bibliography of publications relevant to the subject of this book is included in References and Further Reading.

11. Index.

In writing this book, it is the author's hope that it will prove useful for those involved in both graduate and postgraduate work in research and development and also practising engineers in industry.

Illés DUDÁS

ACKNOWLEDGEMENTS

In the course of the research on which much of the content of this book is based, there were many people who assisted me or contributed directly or indirectly to my work. To them, I would like to give my heartfelt thanks.

While I cannot manage to mention all of those involved, I should like to express my grateful thanks to the following, who helped and encouraged my development in this area of research.

First, from my undergraduate days, when I first became interested in the correlation between research and manufacture, Dr József Molnár, who played an important role in bringing to my attention the lack of precision in the WFMT; in the various phases of my research, Drs Károly Bakondi, István Drahos and Imre Lévai, and also Drs Tibor Bercsey and József Hegyháti of the Technical University of Budapest for their active and helpful collaboration.

My thanks are also due for the help of Professor Dr Zéno Terplán, who was the Head of Department of Machine Elements at the University of Miskolc for several decades, and I am grateful for having been able to consult Professors Friedhelm Lierath (Otto-von-Guericke University of Magdeburg), Boris Alekszeyevich Perepelica (Technical University of Kharkov) and Hans Winter (Technical University of Munich).

I should like to express special thanks to Professor Dr Faydor L. Litvin (University of Illinois, Chicago), whose work helped me in my basic research and who allowed me to consult him personally, and I should like to acknowledge my debt to Dr F. Handschuh (NASA Glenn Research Center), whose critiques in various publications of The American Society of Mechanical Engineers increased the scope of my awareness.

I should also like to acknowledge the help of Tivadar Garamvölgyi and Nándor Less who, during my time at the Machine Factory in Diosgyor, where I was able to access and experiment with a range of production equipment, made it possible for me, by solving constructional and technological tasks, to carry out my research. I also thank János Ankli, who gave me active help during the experimental production.

In the preparation of this book I am grateful, for his technical consultation, to Professor Imre Lévai of the University of Miskolc. Drs Károly Bányai and Gyula Varga read the entire manuscript, my colleagues Drs László Dudás and Gábor Molnár, and Dr Tamás Szirtes, visiting professor at the University of Miskolc, from Ontario, Canada, read individual chapters of the book and gave useful advice. Furthermore, Dr Tibor Csermely helped me in the development of the methods of measurement.

Thanks are due to my colleagues in the Departmental Research Group of the Hungarian Academy of Sciences, who helped with the editing of the text, the running of different programs and the preparation of the various figures and illustrations. They are: András Benyó, János Gyovai, István Fekete, Csaba Karádi, Mihály Horváth, András Molnár, László Sisári, József Szabó, János Szénási and Zsolt Bajáky; Noémi Lengyel and Eleonóra T. Hornyák are to be commended for their careful typing.

Dr István Elinger of the Technical University of Budapest was responsible for the initial translation into English and Illés Szabolcs Dudás collaborated with Penton Press in the final English version.

Finally, I should like to express my grateful thanks to Professors József Cselényi, Pro-rector, and Lajos Besenyei, Rector of the University of Miskolc, for their outstanding encouragement and support.

The following research projects, led by me, provided financial support for the research:

'Optimisation of toothed driving pairs and gearing, development of their mating and their tribology' (OTKA-National Scientific Research Basic Programs – T000655 BME-ME). 1991–4.

'System of conditions for the forming of optimum mating' (OTKA – T019093). 1996–9.

'Complex analysis of machine industrial technologies, regarding mainly the research fields of the production geometry of sophisticated geometrical shapes and computer aided production engineering',

MTA-ME Research Group at the Department of Production Engineering. 1996–8.

'Developing of 3D measuring method by the use of CCD cameras'. Hungarian–Japanese common research project by the support of Monbusho Foundation. 1995–7.

'Development of measuring method by CCD cameras on the field of machine industrial quality assurance' (OTKA 026566). 1998–2001.

Finally, thanks are due to Invest Trade Ltd. (Miskolc) for its financial support in the publication of this book.

Illés Dudás
Department of Production Engineering
University of Miskolc, Hungary

Miskolc
March, 1999

LIST OF SYMBOLS

a, b, c	(mm)	Coordinate displacements from origin O_2 in stationary coordinate system fixed to tool K_o
A_{sz}		Centre distance of grinding stone and worm axes
b_1	(mm)	Axial length of worm
d_1	(mm)	Diameter of worm
d_{a1}	(mm)	Addendum cylinder diameter of the worm
d_{g1}	(mm)	Pitch cylinder diameter of the worm
d_{f1}	(mm)	Root cylinder diameter of the worm
d_s	(mm)	Diameter pitch cylinder of milling cutter
$d_{a1,min}$	(mm)	Minimum addendum diameter of conical worm (spiroid worm)
$d_{a1,max}$	(mm)	Maximum addendum tip diameter of conical worm (spiroid worm)
h_{f1}	(mm)	Dedendum height of the worm tooth
h_{a1}	(mm)	Addendum height of the worm tooth
$i_{2,1}$		Gearing ratio [$i_{2,1} = \varphi_2/\varphi_1$]
K	(mm)	The distance of origin of profile radius from worm centre line
K_0 (x_0, y_0, z_0)		Stationary coordinate system affixed to machine tool
K_{1F} (x_{1F}, y_{1F}, z_{1F})		Rotating coordinate system affixed to helicoidal surface
K_1 (x_1, y_1, z_1)		Coordinate system connected to linear moving table

K_{2F} (x_{2F}, y_{2F}, z_{2F})		Rotating coordinate system fixed to tool
K_2 (x_2, y_2, z_2)		Stationary coordinate system fixed to tool
K_{20} (x_{20}, y_{20}, z_{20})		Coordinate system of the generating curve of a tool of surface of revolution
K_k (x_k, y_k, z_k)		Auxiliary coordinate system
$K_s (\xi, \eta, \zeta)$		Tool coordinate system of generating curve of helicoidal surface
m	(mm)	Axial module
$\underline{M}_{1F, 2F}$		Coordinate transformation matrix (transforms K_{2F} to K_{1F})
$\underline{M}_{2F,1F}$		Coordinate transformation matrix (transforms K_{1F} to K_{2F})
$\underline{M}_{2F,20}$		Coordinate transformation matrix (transforms K_{20} to K_{2F})
$n^{(2)}$		Unit normal vector of tool surface
n_1	(min^{-1})	Number of revolutions of worm
n_2	(min^{-1})	Number of revolutions of worm wheel (spiroid gear)
n_{1F}		Unit normal vector of helicoidal surface in coordinate system K_{1F}
n_{2F}		Unit normal vector of tool surface in coordinate system K_{2F}
$O_0, O_1, O_2, O_{1F}, O_{2F}, O_k$		Origins of coordinate systems related to their subscripts
p		Screw parameter of the helix on worm
p_t		Tangential screw parameter
p_a		Axial screw parameter
p_h		Lead parameter of the face surface on hob
p_r		Radial screw parameter
$P_{1h}, P_{1k}, P_{1s}, P_{1a}$		Kinematic projection matrix for direct method (cylindrical, conical, general)
$P_{2h}, P_{2k}, P_{2s}, P_{2a}$		Kinematic projection matrix, for inverse method (cylindrical, conical, general)
P_x	(mm)	Axial pitch of the worm
P_z	(mm)	Lead of thread
r_D	(mm)	Radius of root circle of worm (convolute)

$\vec{r}_{gsz}$		Position vector of the generating curve of tool surface
$\vec{r}_{2F}$		Position vector of moving point of tool surface
$\vec{r}_t$		Position vector to contact point of tangent sphere with thread surface
r_a	(mm)	Radius of worm base circle
R_{sz}	(mm)	Radius of tool
S_1	(mm)	Tooth thickness of the worm
S_{1F}	(mm)	Tooth thickness of dedendum of the tooth of the worm
S_{2F}	(mm)	Tooth thickness of dedendum of the tooth of the worm gear
x_2		Coefficient of profile displacement
X, Y, Z	(mm)	Coordinates of centre of feeler sphere
x_m, y_m, z_m	(mm)	Coordinates of a measured point
$\Delta x, \Delta y, \Delta z$	(mm)	Coordinates (in three orthogonal directions) of distances between contact point and feeler's centre
$v_{1F}^{(1,2)}$	(m min^{-1})	Velocity vector of helicoid and tool surfaces in the K_{1F} coordinate system
$v_{2F}^{(1,2)}$	(m min^{-1})	Velocity vector of helicoid and tool surfaces in the coordinate system K_{2F}
v_k	(m min^{-1})	Peripheral speed of the worm
z_1		No of starts on the worm
z_2		No of teeth on worm wheel
z_{ax}	(mm)	Axial translation of helicoidal surface to the manufacturing position
α (°)		Forming, tilting angle (in degrees) of tool into profile of helicoidal surface in the characteristic section, eg grinding of involute helicoidal surface using plane face surface wheel
β (°)		Forming angle (in degrees) in the forming plane, which is the height of the radius of base cylinder (spiroid worm)
β_j, β_b (°)		Profile angle (in degrees) profiles on the right and left sides of tooth conical worm
γ	(°)	Lead angle (in degrees) on the worm's references surface

δ_1	(°)	Half-cone angle of the reference cone of conical worm
δ_{ax}	(°)	Tangential angle of the arched profile of the worm on the pitch cylinder
ρ_k	(mm)	Radius of grinding stone in the axial direction
ρ_{ax}	(mm)	Radius of tooth profile of worm having circular profile in axial section
$\left.\begin{array}{l} u \\ \eta \\ \vartheta \\ \vartheta_1 \end{array}\right\}$		Internal parameters of the helicoidal surface (where $\eta = u\,\cos\beta$)
η		Efficiency of the worm gear drive
φ_1	(°)	Angular displacement of a helicoidal surface
φ_2 (°)	(°)	Angular displacement of a tool
$\varphi_{1.opt}$	(°)	Optimal displacement of a worm (place configuration error is minimum)
$\left.\begin{array}{l} \psi \\ y_{20} \end{array}\right\|$		Internal parameters of tool having surface of revolution
ω_1	(s⁻¹)	Angular velocity of worm
ω_2	(s⁻¹)	Angular velocity of tool
ζ, η, ξ		Axes of the coordinate system (K_{sz}) of the tool

1

INTRODUCTION

This book, using data both from the published literature as well as that derived from results of my own research work, aims to discuss general problems of manufacture, the theory of precise geometrical design, the principles involved in inspection, and the problems arising from the use of different types of helicoidal surfaces.

The aims of the book are:

- to develop and to solve generally valid formulae for precise geometric truing of helicoids based on the established and latest results developed in tooth geometry and in kinematic geometry;
- to determine the precise shape of the disc-type grinding stone and to develop tools for shaping;
- to analyse edge tools having regular edge geometry;
- to formulate mathematical equations for the required geometrical and contact conditions;
- to determine checking and inspection methods;
- to create systems for different types of helicoidal surfaces on the basis of their common manufacturing geometry;
- to develop special tools needed in manufacturing, using up-to-date manufacturing systems.

In Figure 1.1 the most frequently used helicoids having cylindrical or conical pitch surfaces are tabulated. The main fields of use are shown to be kinematic and power drives. At the same time it indicates that a generally valid model for helicoidal surfaces with cylindrical and conical pitch surfaces may find application in engineering practice.

Referring to Figure 1.1, the basic fields of application are as follows:

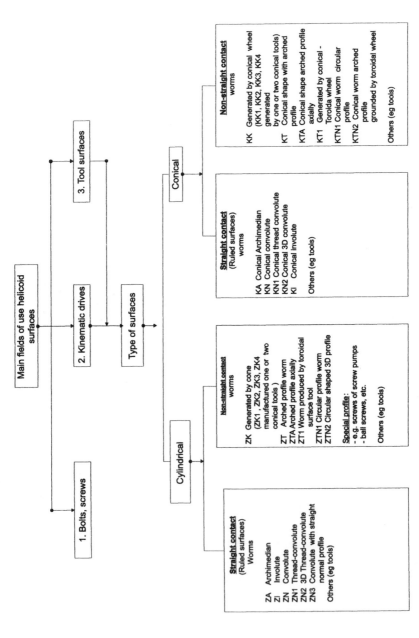

Figure 1.1 *Most frequent applications of helicoidal surfaces (to find any particular subject see table in Figure 1.2)*

- *Thread profiles serve basically as joining elements.* From the point of view of manufacturing geometry, these elements can be manufactured to inferior standards to meet the technical requirements. These are outside the scope of this book.
- *The active surfaces of cylindrical and conical worms (spiroid worms)* used as elements in a kinematic chain can be employed in material handling equipment (cranes, belt conveyors, etc). It should be mentioned that conical helicoids have not been standardized by ISO as yet but may be in the near future.
- *Tool surfaces*, the main and side surfaces of generating, disc-type and shape milling cutters, for thread cutting tools and for surfaces of truing grinding stones.

To design and manufacture these surfaces with precision, it is useful to formulate a generally valid mathematical model that can become the basis for CAD/CAM/CAQ/CIM systems to be developed.

Application of CAD facilitates design of helicoidal surfaces and their generating tools: use of CAM solves the fine truing of grinding wheels using CNC programs; CAQ provides the possibility of 3D-type automatic checking to maintain high quality of product, all within a computer integrated manufacturing (CIM) system.

Traditional machine tools (not controlled by programs) but employing additional automation can be adapted for the manufacture of precise helicoidal surfaces and they can be connected into modern manufacturing systems. In this way, helicoidal surfaces created in manufacturing cells can be manufactured with precise geometry irrespective of the manufacturing scale required.

To apply CAD and CAM to the manufacture of helicoidal surfaces as threads and their mated elements together with their generating tools is especially important in individual manufacture where the number of variants is high and consequently the variety of tools required is high as well.

Applying the work of H. I. Gochman and F. L. Litvin to the field of tooth geometry and theory of gear kinematics, further combining them with the matrix method to transfer coordinate systems into each other, as well as using methods of differential geometry, the problems of manufacturing helicoidal surfaces can be presented in algorithmic form.

The systems developed for design, manufacture and checking were tested in the course of the manufacture of drives and their manufacturing tools.

1.1 CLASSIFICATION OF WORM GEAR DRIVES

From a functional point of view:

(a) *Kinematic drives* can be characterized by adjustable centre distance – shaft angle may incline from 90°. Used in measuring instruments and dividing mechanisms, generally to transfer low power.

The usual ranges of basic dimensions are:

$$1 \leq m \leq 16$$

$$d_2 \text{ up to } 5000 \text{ mm}$$

(b) *Power drives* characterized by fixed centre distance – shaft angle is strict 90°.

The usual ranges of basic dimensions are:

$$1 \leq m \leq 30$$

$$d_1 \text{ up to } 400 \text{ mm, } d_2 \text{ up to } 2000 \text{ mm}$$

From a constructional point of view:

Depending on characteristic shapes of worm and wheel there are three types (see Figure 1.2):

(a) *Cylindrical* (cylindrical worm). Figure 1.2(a). Great Britain, Germany, the former Soviet Union, Hungary.
(b) *Globoid* (worm is globoid). Figures 1.2(b) and 1.2(c). USA, the former Soviet Union.
(c) *Special* (either worm or worm gear has special shape, eg spiroid, USA, Hungary, etc). Figure 1.2.(d).

Classification of cylindrical worms

Of the three types mentioned above, the cylindrical worm is the most widely known and used. Worms manufactured using a straight-edge cutting tool (straight-line contact line is situated on worm tooth surface) are classified as follows:

Cylindrical worms with ruled surface

- Archimedian worm (ZA) having line axial section.
- Thread-convolute worm (ZN1) having lines normal to thread surface.

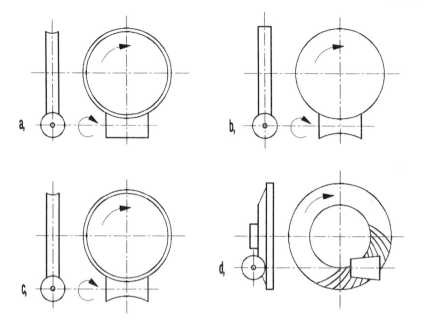

Figure 1.2 _Classification of worm gear drives from constructional point of view:_

    (a) cylindrical worm (cylindrical worm–globoid worm wheel);
    (b) globoid worm (globoid worm–cylindrical worm wheel);
    (c) globoid worm (globoid worm–globoid worm wheel);
    (d) conical worm (spiroid worm–face wheel).

- 3D thread-convolute worm (ZN2) having lines normal to tooth space;
- Involute worm (ZI) having lines fitted on plane tangential to base cylinder;
- Duplex worm (ZD) having lines on different leads on different sides of worm.

Non-ruled surface cylindrical worms

Worms are manufactured using a straight-edge cutting tool, but a straight line on a worm surface can be fitted.

- Single cone worm: ZK1 end milling cutter or pin grinding wheel;
- Double cone worm: disc milling cutter ZK2 or double pin grinding wheel;

Arched profile worms

- Worm profile in axial section arched: ZTA, the circular profile, is in the axial section of worm;
- Thread worm with arched profile: ZTN1, the circular profile, is in the plane normal to tooth surface;
- 3D thread worm with arched profile: ZTN2, the circular profile, is in the plane normal to thread path along the middle of tooth space;
- Worm having arched profile produced by double circular disc milling cutter: ZT1, the worm profile, is determined by the disc milling cutter with double circular profile inclined to tooth space by δ lead angle (in degrees) on the worm's reference surface.

Standardized types of worm are differentiated by the differences in tooth surfaces. Table 1.1 summarizes worm types identifying the principal section and the geometric shape of generator.

Table 1.1 *Classification of worms depending on the mother tool – unit normal vector of helicoidal surface*

Important section	Generator	Generating Tool (Mother Tool)			
		Plane	Rack	Single Cone	Double Cone
Axial	Line	ZA			
	Arc	ZTA			
Normal to tooth	Line	ZN1			
	Arc	ZTN1			
Normal to tooth space	Line	ZN2		ZK1	ZK2
	Arc	ZTN2			ZTK
Normal to cylinder	Line		ZI		
	Arc				

2

A SHORT HISTORY AND REVIEW OF THE LITERATURE

2.1 A SHORT HISTORY OF THE WORM GEAR DRIVE

Our short history begins with the First Punic Wars, which started in 264 BC and lasted 23 years.

During the first year of Punic War I, Hieron II ascended to the throne of Syracuse and then concluded an alliance with Rome. But Heiron, who never really trusted Rome, initiated a fleet-building programme which included a warship of a size never seen before.

In *The Ships*, by H. W. Van Loon, it is mentioned that the then average size of a ship was 20–30 tons, so it is very likely that Hieron's giant warship could not have exceeded 40–50 tons.

Contemporary shipyards were not able to launch ships of such weight, so Hieron enlisted the assistance of Archimedes, one of the greatest scientific figures of the time.

Archimedes, at Hieron's request, developed a revolutionary crane that made it possible, with the help of a few slaves, to launch the giant ship.

It was at the launching that Archimedes is supposed to have uttered his famous dictum 'Give me a fixed point and I will remove the complete world from its corner points'.

Archimedes called his crane the 'barulkon', and there is little doubt that in the course of its development in the years 232–31 BC he was the first originator of the worm drive (see Figure 2.1).

In the following centuries, the use of the worm gear drive became widespread throughout the then known world.

Figure 2.1 *The Archimedian barulkon (Reuleaux)*

Archimedes, as was then customary, lodged a description of his invention in the Alexandrian Library. From this, the Alexandrian Heron, the other leading technocrat of the ancient world, learned about it, and in about AD 120, that is 350 years later, wrote a book on the barulkon.

Later, during the third century AD, Pappus gave a detailed description of the barulkon in his summarizing work; this consisted of four pairs of gear trains with a worm gear drive added. Pappus mentioned Archimedes as the original inventor, and Reuleaux, a German engineer, used Heron's description to reconstruct and make a drawing of the barulkon (see Figure 2.1).

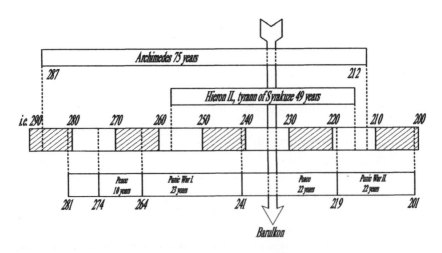

Figure 2.2 *History of the invention of the worm gear drive*

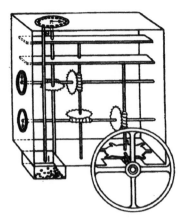

Figure 2.3 *Vitrinius' taximeter (Diels)*

The next author mentioning the worm gear drive was the Roman architect Vitrinius. In his book *De Architectura*, published between the years 30–16 BC, there is a description of the *hodomate*. The Romans' rented passenger vehicles were equipped with a 'hodometer'. In this device small balls were allowed to fall one-by-one into a drawer, each denoting the fulfilment of a mile.

This invention provided the first recorded taximeter (see Figure 2.3).

The first technically significant worm gear drawings were made by Leonardo da Vinci (1452–1519). These drawings were found among his sketches and notes. Surprisingly, not only drawings of worms and worm gears but also of globoid type worms are to be found among them (Figure 2.4a).

So Leonardo had known of an asymmetrical worm (Figure 2.4b) that could be mated with a sprocket wheel and he was aware of the crossed helical gears substituting the worm with more than one start (Figure 2.4c). He designed control drives too, where the driving element was the worm gear driving the worm (Figure 2.4d), but these have not been used in practice.

Far more interestingly, from the sketches it is evident that the principle of self-locking had already been known to Leonardo.

In the case of the small pitch worm (Figure 2.4e), he stated: 'This is the best type of worm gear drive, as the worm gear would never be able to drive the worm'. He also stated, referring to high-pitch worms (Figure 2.4f): 'The simple high-pitch worm can be easily driven, easier than any other type of drive'.

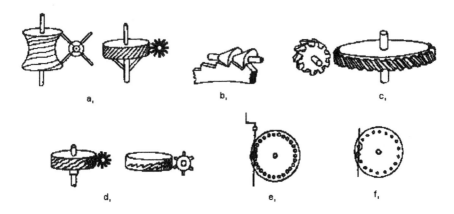

Figure 2.4 *Sketches of worm gear drives made by Leonardo da Vinci*

He had come to the conclusion that in drives having skew axes, among them worm gear drives, two different types of contact, the sliding and the rolling ones, come into effect.

In the four centuries from Leonardo's age to ours, in all the technical books published, worm gear drives can be found everywhere when a high gear ratio should be used.

From the time that the use of the steam engine had become widespread and step-by-step machine tools had become increasingly used, that is starting at the second half of the eighteenth century, the worm gear drive became a popular machine element.

In an early application, the Boulton and Watt factory in England manufactured for this purpose 15in diameter (380 mm) big worm driving gears with wooden teeth, but eventually, as the mechanical demands were increased, the wooden teeth were replaced by iron ones.

In the two decades between 1880–1900 the electric motor became widely used. Unfortunately, owing to its relatively high shaft speed, the electric motor could not use systems in the technology of the period. The available gears were not suitable. Worm gear drives, because of their high gearing ratios, were not suitable for high-input shaft speeds. These drives became significantly overheated, wear was high and the technique of lubrication was poor for low-input shaft speeds. Grease lubricant had to be used because there was no housing to contain the oil. It is well known that dissimilar materials operate more effectively in tribological pairs.

The development from slow worm gear drives to high shaft speed modern ones started with the introduction of closed housings to facilitate oil lubrication.

It became clear that, to reduce frictional losses, worms and worm gears should be manufactured from different engineering materials, namely the worm from steel, and the worm gear from cast iron or from bronze.

The history of tooth generation technologies shows a very interesting relationship between theory and practice. The theory was in place at least 200 years before the generation of teeth as spur gears.

It appears that there were very few theoreticians capable of dealing with worm gear drives after Archimedes, and very few technicians capable of applying the known theory.

The fact is that for the centuries after Archimedes the need did not arise for such geared systems until the invention of the electric motor.

The theoretical investigations needed to dimension worm gear drives made rapid strides with the development of the electric motor. The most outstanding workers who dealt with the question were Bach and Stribeck. At that time the dimensioning of spur gears was treated as if the tooth was a loaded cantilever beam. This was suggested first by Tredgold, an English engineer in 1882, who formulated $P = kbt$, where k expresses the sum of practical experiences, as a coefficient.

Work then commenced on the precise geometry of the worm, the method developed being a purely theoretical one.

These geometrical investigations of the worm led to the concept of the _involute worm_. Previously, all the worms manufactured using a straight edge tool had been called involute worms. Thanks to the precise work carried out the term now refers only to a worm manufactured under strictly circumscribed conditions and their results served well the practice of worm grinding too.

The geometrical view held sway until recently, but is now being replaced by the functional view, which is better suited to gear manufacture. The new concept, published by Szeniczei (1957) investigates geometry of worm gear drives from a functional point of view independent of whether the worm has involute profile or not. The Wildhaber theory (1926) was the ruling geometrical view, mainly emanating from the German workers; they thought that the involute worm, having the equivalent geometry of helical cylindrical teething, would automatically solve the problems of worm gear drive completely. So completely did the 'cult of the involute' occupy for decades the German technologists, that all other branches of research, eg the investigation of globoid drives, was completely neglected.

The role of globoid worm tool is very interesting. As has already been mentioned, although among Leonardo's notes and sketches the globoid worm was noted, no further data surfaced. But later, when the disadvantages of the cylindrical worm driven by electric motor became clear, the globoid principle was immediately taken up in the hope that the higher load-carrying capacity would provide better conditions of engagement.

British and especially American workers developed the new technology step-by-step and found it good enough to use in practice, obtaining even better results than with cylindrical worms. It is now generally used throughout Great Britain and the USA.

Hindley manufactured the first globoid worm gear drive (Buckingham, 1949) in 1765. The first globoid drives were manufactured by Hughes and Philips in America in 1873 and by Crozef-Fourneyron in France in 1884. Wildhaber used the globoid worm mated with a cylindrical wheel having a ruled surface first in 1922 for precise adjustment of scales in gauges. Later, this type of drive was developed for application in power drives.

Samuel Cone (The Michigan Tool Co) can be regarded as the real master of globoid drives, reporting his first patent in 1932, which was followed by several others. Previously, the so-called *double enveloped cone* type, a globoid drive, was known from the year 1924.

The literature of worm gear drives has been sparse. Most authors had little connection with manufacturing technologies and therefore applications and practical experience were not being given sufficient consideration.

As a result of technological progress, the use of cylindrical worm drives became widespread throughout Europe – in England, Germany, Russia and in Hungary too.

The use of globoid worm gear drives became common first of all in the USA and in the Soviet Union but their use was investigated in Germany and in Hungary too. Spiroid worm gear drives, a category of special drives, had been patented in the USA but also achieved success in Russia, in Germany, in Bulgaria and in Hungary.

The literature relating to the meaningful development of worm gear drives that happened at the end of the 19th and during the 20th century is reviewed in the following section.

2.2 DEVELOPMENT OF TOOTH CUTTING THEORY FOR SPATIAL DRIVES

The investigation of the tooth cutting theory for flank gears, and the systematic collection of results, lasted for several decades. The first papers were published in the mid-19th century and dealt basically with two areas of tooth-generation theory, the conditions for teeth meshing and the manufacture of them (Gochman, 1886; Olivier, 1842; Reuleaux, 1982). The Frenchman Olivier, whose investigations were outstanding over a long period, in his book, published in 1842, separated the theory of meshing from analytical and enveloping (geometric) methods. He regarded the theory of meshing as purely a question of 'descriptive geometry' while the Russian Gochman stated that 'the theory of tooth-cutting is a special field of mathematics where the investigator, in contrast with other fields of mathematics, should progress step-by-step searching safe points at each step'.

Both views were too general and neither can claim exclusive right to state a generalized spatial tooth-generation theory, the basis of which was laid down in the works of Olivier (1842) and Gochman (1886).

It was Gochman who first applied an analytical model for the investigation of meshing spatial surfaces and formulated the mathematical description of wrapping surfaces.

The theory of tooth generation draws on many different areas like differential geometry, manufacturing, design, metrology and computer technology. The development of tooth generation using computer techniques has transformed it into a modern theory and it has now been expanded for direct use in industry. In our day it can be regarded as a specialist area of technical science.

The work of Distelli (1904, 1908), Stübler (1911, 1922), Altmann (1937) and Crain (1907), published in the first years of this century, should also be mentioned. They all achieved significant results by applying the methods of descriptive geometry to the development of tooth-generation theory. The phenomenon of vector-twist was first mentioned by R. Ball in 1900 and Distelli was one of the pioneers who applied, in his work published in 1904, general screw motion to describe mated gears having skew axes. Formulation of the driving-twist or screw-axoid made it possible to describe in a simple and clear way the process of manufacturing working surfaces of teeth mating along a line attached to each other. His works (Distelli, 1904, 1908) dealt with ruled surfaces, the simplest ones.

The names of Willis (1841), Buckingham (1963), Wildhaber and Dudley (1943, 1954, 1961, 1962, 1984, 1991) are also to be noted.

It was Willis (1841) who suggested the law governing contact of plane curves.

Generalizing Distelli's work (1904, 1908), Wildhaber (1926, 1948) successfully mated theory and practice, and by applying the kinematic method to the theory of gear drives achieved significant results. His findings have been subsequently supported by Capelle (1949).

Applying mathematical methods, several researchers investigated the possibilities of determining mathematically the surface mated to another for known centre lines and angular speed ratios. Difficulties had often arisen in formulating and examining either analytically or numerically these complicated equations.

Significant research was carried out by Hoschek (1965) on mated elements having closed wrapping surfaces.

Based on Grüss' results (1951), Müller (1955, 1959) developed a method of determining a wrapping curve for plane teeth. His mathematical equations could be applied to a few types of spatial drive only. The analytical and geometrical methods developed then are still used for investigating spatial teethed drives. It gradually became apparent to researchers dealing with theoretical problems of gear drives that application of the so-called kinematic method had the effect of simplifying research.

Based on this method, Litvin and other outstanding representatives of the Soviet school dealing with theory of gear drives, for example Kolcsin (1949, 1968), Krivenko (1967), Litvin (1962, 1968, 1972), worked out suitable and efficient methods to determine equations and criteria for mating and contact conditions, for characteristics of curvature of contact surfaces and for the phenomena of interference.

Further researchers worthy of mention are Bär (1977, 1996), Ortleb (1971), Wittig (1966), Jauch (1960) (work on surfaces of worms) as well as Dyson (1969) (general theory of teeth generation), Zalgaller (1975) (who developed the theory of wrapping surfaces) and Buckingham (1960) (research in the field of involute worm gear drives).

Research work on the geometry of teeth generation, ie the working out, systematizing and analysis of machining processes, have found a new impetus in the last decades. Weinhold (1963), Kienzle (1956) and Perepelica (1981) have thrown new light on these problems.

Outstanding Hungarian scientists in this field, producing valuable results, were Szeniczei (1957, 1961), Tajnaföi (1965), Magyar (1960), Drahos (1958, 1967, 1973, 1981, 1987), Lévai 1965,1980), Bercsey (1977, 1999) and Drobni (1967, 1968, 1969).

It was Szeniczei (1961) who first conceived the idea of *conjugated surfaces* (mutually wrapping surfaces). Then it was Magyar (1960) who first enlightened the problems of meshing with worm surfaces. Tajnaföi (1969) determined and systematized the generally valid theoretical bases of teething technology and the principles of the movement-generating characteristics of machine tools. Drahos (1958) investigated the geometry of several tools, helicoid surfaces and especially hypoid bevel gears and, in his analyses in the field of machining geometry, enriched the theory (Drahos, 1987). Then Lévai (1965, 1980) dealt with several problems associated with spatial drives. He was investigating the theory of teething in drives between skew axes with ruled surfaces transferring variable speed to the movement. Later, he dealt with basic problems in the design of hypoid drives.

Bercsey (1977) analysed the possibility of application of the kinematical method and investigated partly the meshing conditions of a globoid worm, manufactured with ruled surface, with a hyperbolic gear applying the kinematic method in practice, and partly the toroid drives. He proved possible the application of this method to these types of drives, making it possible to analyse, using similar methods, the meshing conditions of other spatial drives.

As a version of worm gear drives using involute teething, Bilz (1976) in Germany, developed an element of the cylindrical gear, globoid worm gear drive family, the 'TU–ME' globoid drive. The theoretical investigation of this drive was carried out by Drahos (1981).

Drobni (1967) dealt with globoid worm gear drives and worked out a globoid worm gear drive with ground surfaces. In this work he verified that it is not necessary for a worm to be manufactured with a trapezoidal cutting tool edge situated in the axial section of the worm to which an undercutted gear belongs. But it can be manufactured by generating it using a transferring generator surface or by indirect movement generation (using the theory of conjugated teeth) and so the worm becomes suitable for grinding the worm wheel teeth, is axially not undercut and there is no need for separate correction for the worm body. Siposs (1977) also contributed to this work.

2.3 CYLINDRICAL WORM SURFACES

The cylindrical worm surface can take the form of a ruled surface (with line generator) or non-ruled surface (no line generator).

The ruled worm surface can be Archimedian, having a line edge in axial section or a convolute one having line intersection in section with a plane parallel to the worm axis. So the involute worm surface is a special case of convolute worm surface having a line edge section in a plane parallel to the worm axis and tangential to the base cylinder of the involute. A class of non-ruled worm surfaces is the ZK type worms characterized by a line as the meridian curve of the generating tool (Litvin, 1968, 1972; Maros, Killmann and Rohonyi, 1970; Niemann and Winter, 1965, 1983). The position of the line edge tool relative to the helicoidal surface will determine the direct sub-class within ZK type worms ZK1, ZK2, ZK3 and ZK4 (ISO701 (1976) and ISO1122/1 (1983) MSZ (Hungarian Standard)). Kinematic relations between the generating tool and the machined surface determine the profile of the helicoid surface.

2.3.1 Helicoidal surfaces having arched profile

One the most modern types of cylindrical helicoidal surfaces is the worm generated using a circular profile tool. Depending on the kinematical conditions between worm and tool, the circular profile can appear directly on the worm active surface (in axial or normal section (Krivenko, 1967), perhaps in a plane section parallel to the worm axis) but in certain cases (for example, manufacturing with a disc-like tool having circular axial section) (Flender and Bocholt, 0000; Patentschrift, Deutsches Patentamt, Nos 905444 47h 3/855527 27h) it does not necessarily happen.

Contact surfaces between worms having ruled surfaces (Archimedian, convolute, involute types) and mated worm wheels do not allow the formation of a continuous, high pressure-bearing oil film. It is best to build up an oil film between mated surfaces so that the direction of the relative velocity of the drive faces into the direction normal to the common contact curve or very close to it. More advantageous conditions exist for circular profile worms. The David Brown Company first manufactured a worm–worm wheel drive with this profile. The axial section of their worm was convex arched while the profile of the mated worm wheel had a concave shape in axial section.

Niemann determined, after detailed investigation of lubricating conditions, that circular profile worms are the best from this point of view (Niemann, 1956, 1965; Niemann and Heyer, 1953; Niemann and Weber, 1942, 1954; Niemann and Winter, 1983).

To clarify the conditions, based on the literature (Niemann, 1965), the momentary contact curves are presented in Figure 2.5 for ruled surface involute and curved profile (CAVEX) cylindrical worms. The vector of the sliding velocity (relative velocity) of worm is nearly parallel to these contact curves (see lines 1, 2, 3 in Figure 2.5a). More precisely, the vector component parallel to tangents of these curves $\vec{V}_k$ is significant. Worms having a curved profile meet these conditions far better.

Based on Niemann's tests and patent, the German company Flender produced the CAVEX type worm gear drive (Niemann and Weber, 1942; Patentschrift, Deutsches Patentamt, Nos 905444 47h 3/ 855527 27h), for which the contact curves and relative positions of velocity components are shown in Figure 2.5b. From the figure it can be seen that at the oblique chosen contact point the tangent of momentary contact curve is nearly perpendicular to the relative velocity vector.

Thanks to the wedge shape clearance between the teeth, facing into the direction of relative velocity, a continuous oil film with load carrying capacity is formed, providing pure hydrodynamic lubrication between the teeth of the driving and driven elements.

In Figure 2.5 $\vec{V}_k$ denotes the peripheral velocity of the worm, which for the point of contact fits into the plane of the drawing and is the component of the relative velocity as well. The velocity $\vec{v}$, being perpendicular to the contact line, is the entrainment velocity of the momentary contact line. To obtain advantageous conditions at contact with reasonably high hydrodynamic pressure, this velocity should be as high as possible .

As the figure shows, it is usual for two or three teeth of the worm gear to be simultaneously engaged. The contact curve at a given tooth is wandering axially through lines marked 1, 2, 3 from step into engagement to leave the contact zone while it revolves along acting active surfaces.

A further advantage of mated elements having curved profile is that the radii of mating flank surfaces are situated on the same side of the contact point common tangent, that is, concave and convex surfaces are in contact, generating a relatively small Hertz stress. This is why the load-carrying capacity of a worm gear drive having

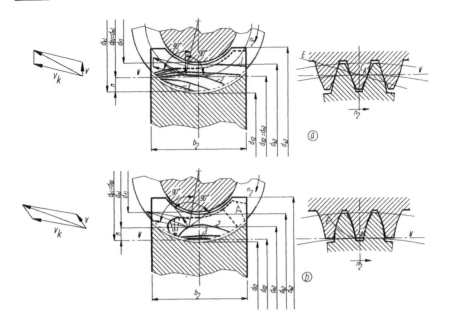

Figure 2.5 *Teeth connection and contact lines of tooth flanks Contact line in axial section is E; a involute worm gear drive; b curved profile (CAVEX) worm gear drive; basic parameters of drive are equal (Niemann, 1965)*

curved profile is significantly higher than a cylindrical worm gear drive with ruled surface and of equal basic size.

As a consequence of smaller contact pressure, the load-carrying oil film can form easier.

This type of drive, in the case of a non-localized contact area, is extremely sensitive to heat expansion, to mechanical elastic deformation and precision of assembly.

As a result of arched profile teething the tooth shape and the suitable positioning of the centre of curvature of tooth flank (the position of engagement line), the dedendum tooth thicknesses, both on the worm $\overline{S}_{1F}$ and worm gear $\overline{S}_{2F}$, are significantly wider. Dedendum tooth thicknesses both of worms and worm wheels manufactured with ruled surfaces are smaller.

Summarizing the principle of tooth shaping as illustrated in Figure 2.6, the teeth of worms have concave profiles instead of line edge or convex; and the pitch line (d_{g1}) of worms is situated in the vicinity of the addendum cylinder or the mean of tooth height is

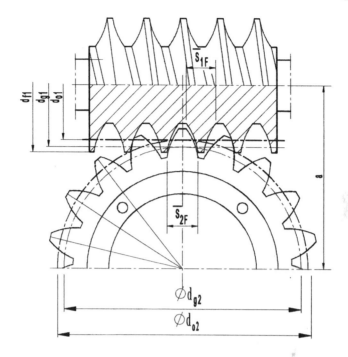

Figure 2.6 *Principle of tooth shape, position of pitch line*

outside the reference cylinder (d_{01}) owing to the fact that the value of the profile displacement coefficient (x_2) is in the range

$$0.8 \leq x_2 \; 1.5$$

As quoted in Muller (1955) the efficiency of worm gear drives (in case of slow-down drives) is given by:

$$\eta = \frac{tg\,\gamma_0}{tg\,(\gamma_0 + \rho)} \qquad (2.1)$$

Where γ_o is the lead angle of worm on reference cylinder, and ρ is half cone angle of friction ($\mu = \tan\rho$).

The correlation is valid in the case of an arbitrary type of worm.

The efficiency as calculated increases sharply with an increase in lead angle γ_o of up to 15°, but above $\gamma_o > 15°$ the increase slows down (see Figure 2.7) especially when $\mu = \tan\rho$ is small.

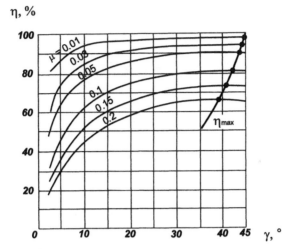

Figure 2.7 *Efficiency of worm gear drive as a function of lead angle γ_o and friction coefficient μ*

Based on equation (2.1) it seems that the efficiency of a worm gear drive is independent of power transmitted, as neither the sliding velocity nor the contact pressure on the mated teeth are involved in the equation. But the friction coefficient is in practice influenced by the surface roughness of the mated surfaces, the temperature and pressure of lubricant and the value of sliding velocity.

The efficiency of worm gear drives, as stated in Muller (1955), can be increased by increasing the sliding velocity (increasing shaft speed of worm), using a higher lead angle for the worm (reducing gear ratio), and increasing the geometric sizes of the drive (characteristic value is the centre distance).

Judging from the current author's personal experience, efficiency owing to constructional factors can be further increased, or perhaps decreased, by use of precision manufacturing technology. These factors in manufacturing technology determine the surface roughness and the precision of the geometrical surfaces of the elements. From the point of view of efficiency and of the expected service life of worm gear drives, the tooth shape and the quality of flank surfaces play extremely important roles.

2.3.1.1 Development in manufacturing worms having arched profile

When preparing manufacturing process plans, the principles of precision assurance should always be taken into account. According to this principle all technologies should be able to correct errors created in the earlier stages of the manufacturing process, depending on the system of machine tool (Bálint, 1961). Consequently in manufacturing processes as a whole it is important that precision can be obtained (in size, geometric shape, and relative positioning) in the rough and semi-finishing processes prior to fine finishing (Molnár, 1969).

These days, worms used in high power-transmitting drives by manufacturers applying up-to-date technologies are always finally machined by grinding.

As a consequence of the technology now being applied, the precision of the lead or the pitch is increasing with improvement in surface roughness. Owing to the limited precision in tool shape and the method of manufacture, manufacturing faults occur with continual use, the profile becoming deformed (the precision of tooth shape getting worse) and consequently both the kinematic and load-carrying capacities of the drive are reduced.

The first method worthy of mention was worked out and patented by Niemann in Germany for circular profile worm gear drives (Olivier, 1842). The essence of this method is the toroidal shape disc-like grinding wheel. The grinding wheel used in its axial section has an arc profile, which corresponds to the profile worm in normal section.

The grinding wheel in its axial section is prepared with a circular profile of radius (ρ_k), which is nearly equal to the reference cylinder radius of the worm $(\rho_k \cong d_{o1}/2)$. The axes of grinding wheel and workpiece are skew lines with angle of inclination equal to the lead angle of worm γ_o. The normal transversal line of axes is just situated in the symmetry line of the tooth space in its normal section (Drobni, 1968).

The precise setting of a normal transversal plane for grinding is possible on a machine tool where the grinding wheel can be shifted along its own axis (eg Klingelnberg).

Previously, reduced precision in the shape of the manufactured surface required adjustment of the grinding wheel but nowadays it can be corrected using microprocessor-controlled equipment to meet practical requirements (Predki, 1986).

The centre distance of tool and worm centre lines A_{sz}, according to Figure 2.8 is:

$$A_{sz} = K + r_{sz} - h_{sz} \qquad (2.2)$$

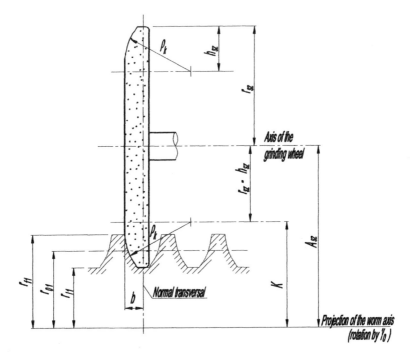

Figure 2.8 *Grinding wheel setting for Niemann grinding (Patentschrift, Deutsches Patentamt, Nos 905444 47h 3/8555 27 27h)*

where:
K is the shortest distance between skew lines centre-line of worm and the line fitted on centre line of the rounding circle of grinding wheel profile and parallel to centre line of grinding wheel,
r_{sz} is the external radius of tool (grinding wheel),
h_{sz} is the distance $K - r_{al}$ between the centre of the rounding circle of the grinding wheel and the addendum cylinder of worm,
ρ_k is the radius of the tooth arch in the projected plane of the grinding wheel where $\rho_k \approx \rho_{pn}$.
The dimension b that determines the position of the grinding wheel, based on Figure 2.9, can be written:

$$b = b_1 + b_2 \tag{2.3}$$

where:

$$b_1 = \rho_k - \sqrt{\rho_k^2 - (K - r_{01})^2} \tag{2.4}$$

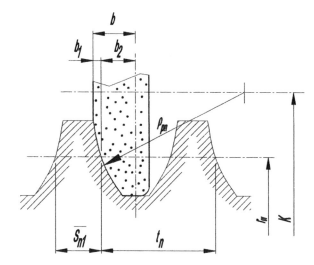

Figure 2.9 _Determination of the position of grinding wheel_

$$b_2 = \frac{P_n - \overline{S}_{n1} - \overline{T}_{S1}}{2} \qquad (2.5)$$

$\overline{S}_{n1}$ is the worm tooth thickness, and $\overline{T}_{s1}$ is the allowable deviation of tooth chord size.

Substituting (2.4) and (2.5) into (2.3) we obtain:

$$b = \frac{t_n - \overline{S}_{n1} - T_{S1}}{2} + \rho_k - \sqrt{\rho_k^2 - (K - r_{01})^2} \qquad (2.6)$$

The disadvantage of this grinding method is that the periodic re-profiling of the grinding wheel causes profile changes. These changes during manufacture ensue, because contact between tool and worm surfaces is a spatial curve, having a character which itself changes owing to wear. The necessary re-profiling of grinding wheel therefore causes changes in its diameter (Niemann and Weber, 1954). As a result of this the precision of shape suffers loss.

The second type of grinding method was worked out and patented by Litvin (Litvin, 1968). In this method, a positioned grinding wheel is used in a special way (Litvin, 1968; Russian Patent No. 139.531). The contact curve between surfaces of grinding wheel and worm is in this case a plane curve equivalent to an axial section profile of the grinding wheel. This method of grinding is based on the fact that two contact axes exist during machining with disc-type tool helicoidal surfaces. The contact axes are lines intersected by the series of normals belonging to different points of the worm contact line. One of these contact axes is the centre line of the grinding wheel, the other one is situated at a certain distance from the centre line of the worm and it intersects the normal transversal line of the axes of worm and tool (Litvin, 1972). The result is an *Archimedian hollow surface* (Drahos, 1966).

$$K = \frac{m\frac{z_1}{8}}{\tan \gamma_0} \tag{2.7}$$

and

$$\delta = \arctan \frac{m\frac{z_1}{8}}{A_{sz}} \tag{2.8}$$

where:
γ_o is inclination of axes of tool and worm,
A_{sz} is centre distance of tool and worm,
m is module in normal section,
z_1 is teeth number of worm (number of starting).

The tool for grinding the worm should be adjusted so that the centre A for profiles labelled 1–2 should fit on axis II–II as in Figure 2.10. This fit occurs if the parameters of the grinding wheel, K and δ, fulfil the requirements of equations (2.7) and (2.8). Naturally, when centre A fits on contact axis II–II then the contact line of tool and worm is itself the profile 1–2 because the normal lines belonging to these surfaces intersect axis II–II (at point A), as well as axis I–I, the axis of the tool.

The verification of this method can be found in Litvin (1972), but the following factors should be treated with care during manufacture.

When re-sharpening the worm tool, the grinding wheel meets the requirement of lead angle γ_o, in the projected plane. The translation of the dressing diamond is brought into conjunction with the grinding wheel to define its profile. This process causes a change in distance and this is why the displacement of the equipment towards the grinding wheel requires the same displacement of wheel axis

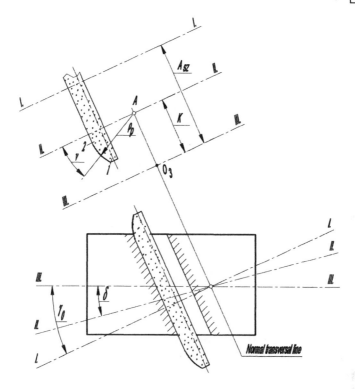

Figure 2.10 _Principle of Litvin grinding method (Litvin, 1968)_

towards the workpiece. It can be stated that precise grinding can only be achieved using a machine tool suitable for adjusting the grinding wheel axially to a significant extent.

There are references in the literature to a number of workers concerned with circular profile worms, including Krivenko in the Soviet Union (Krivenko, 1967) and Kornberger in Poland (Kornberger, 1962).

The number of publications concerning circular profile worm gear drives is small, especially in Hungary. It is easy to understand authors' reticence with several patents protecting manufacturing processes.

In Hungary there were Drobni (1968) and Drahos (1966), who dealt with the construction of Litvin-type contact surfaces, and regarding the checking of this type of drive, Bányai (1977) is worthy of mention. Standardization in Hungary has also showed progress in this field.

The author of this book has taken an active part in the creation of formation standards (KGSZ 07–5501–75, KGSZ 07–5502–75, KGSZ 07–5503–76, MSZ 7490/4–82).

2.3.2 Cylindrical worm gear drives with ruled surfaces

Prior to the 20th century, several researchers, beginning with, for example, Distelli (1904, 1908) and Stübler (1911), dealt with worm drives having ruled surfaces. In recent decades, research workers from Europe, USA and Asia dealt with and obtained results in the field of geometrical shape, production, and checking of these drives (Bar, 1977; Kienzle, 1956; Crain, 1907; Csibi, 1990; Drobni and Szarka, 1969; Dudás, 1976; Höhn, 1966; Kacalak and Lewkowitz, 1984; Litvin, 1968; Maros and Killmann, 1970; Munro, 1991a; Niemann and Weber, 1954; Niemann and Winter, 1983; Ortleb, 1971; Predki, 1986; Umezawa *et al*, 1991; Varga, 1961; Winter and Wilkesmann, 1975; Wittig, 1966).

Previously, helicoids having ruled surfaces were developed in Hungary following World War II when the need for intensive development in this field led to the work initiated by Szeniczei. His book *Worm Gear Drives*, published in 1957, can be regarded as a pioneering work and persuaded several young scientists to work in this field, resulting in a series of interesting publications. Erney (1983) has summarized the results of the Hungarian research work. It was Magyar (1960) who solved the theory and problems of manufacturing involute and convolute helicoids .

At the Diósgyör Machinery Works, Varga dealt with convolute helicoids and produced significant results in this field (Varga, 1961). Series of publications by Tajnafói (1965), Drahos (1967, 1973, 1981, 1987), Drobni and Szarka (1969) illuminated the different areas of the subject. Tajnafói's dissertation (1965), first in this field and in Hungary, made clear the basic principles of movement projections as being closely related to gear teeth generation, and he also pointed out the technological reasons for undercutting. He determined all the necessary and sufficient conditions needed to avoid any kind of undercutting.

2.4 CONICAL HELICOID SURFACES

Conical helicoid surfaces are most frequently used as operating surfaces of conical worms.

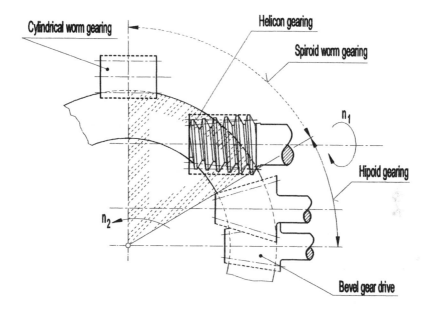

Figure 2.11 _Types of drives classified by relative position of axes_

Energy losses can be significantly reduced in drive systems using modern driving elements of high load-carrying capacity owing to their advantageous hydrodynamic conditions. From the point of view of power loss, the use of tooth shapes having good contact conditions is advantageous. This statement is valid for all types of drives (Hegyháti, 1985, 1988).

Onc of the less known types of toothed drives suitable for torque transfer between orthogonal skew axes, and having significant load carrying capacity, is the spiroid drive. Bohle, at the Illinois Tool Works (USA), in the first publication concerning spiroid drives (Bohle, 1956), stated that it is advantageous for gearing ratios to be at about the limit for bevel gear drives and worm gear drives. This fact is presented in Figure 2.11. The mating elements are a flank gear and, in the general case, a conical worm. When the half cone angle on the worm (δ) is zero, then a cylindrical worm mated with flank wheel is created. Generally, this drive is termed in the literature an _helicon drive_, but the Soviet Standard GOSZT 22850–77 terms it a _spiroid drive_. Bohle (1956) did not refer to numerical data but discussed instead problems of manufacturing technology, the field of its application and operational experience. Characteristic gearing ratios in one step for spiroid drives is i = 10–110, but with specifically chosen characteristics, drives for kinematic tasks with small modules i = 359 has already been realized.

The analysis of the characteristics of spiroid drives in several countries followed the publication of Bohle (1956).

Beside its analysis of contact conditions, it was important in solving problems of manufacture because only reliable, efficient teeth-generating technologies can guarantee the advantageous basic parameters of the geometry determined by theoretical investigation, and produce proper geometric contact.

Along with development of manufacturing technology, Saari (1956) analysed kinematic conditions for spiroid drives.

Results of the research on and operational data for spiroid drives manufactured at the Illinois Tool Works were published in Dudley (1962). The tabulations provided make it possible for designers to compare load-carrying capacity, efficiency, range of gearing ratio, volume required etc with other drives. This handbook is regarded as the basic source concerning spiroid drives. From it we may conclude that load-carrying capacity and possible gearing ratios of spiroid drives are similar to those of hypoid and high power-transmitting worm gear drives while the specific volume required to transmit power is less (Figure 2.11) (Hegyháti, 1985).

Development of spiroid drives in the Soviet Union started in about 1960. In 1977 a standard (GOSZT 22850–77) determining terminology and definitions was prepared. The initial research work dealt with Archimedian (Golubkow, 1959; Nelson, 1961) and then involute worms having ruled spiroid surfaces (Dudley, 1962; Gansin, 1970, 1972; Georgiew and Goldfarb, 1972) as well as their kinematic problems together with their manufacturing technology and operating conditions (Dudley, 1962; Georgiew and Goldfarb, 1972, 1974; Goldfarb and Spiridonov, 1996).

The Bulgarians Abadziew and Minkow (Abadziew and Minkow, 1981; Minkow, 1986) investigated the tooth geometry of spiroid drives. Their work analyses the details of kinematic and geometric problems of spiroid drives with ruled surface. Several researchers (Lange, 1967; Schwägerl, 1967) have tried to compare spiroid and other types of drive (Kolchin, 1968). There are numerous questions to answer in this field, especially the quantitative analysis of contact conditions. The research studies mentioned make it possible to determine the main sizes of drive together with geometric data for teething, but to evaluate conditions of contact further investigations are required. Research activities carried out on spiroid drives in

Hungary by Hegyháti (1985, 1988) from TUB (Technical University of Budapest) Lévai (1980), and Dudás, Cser and Berta (1998), and the manufacturing geometry and tools required were investigated by the author of this book (Dudás, 1986a, 1986b, 1986d). All these workers were from Miskolc University. Their efforts and results make it possible to compare spiroid drives with cylindrical worm gear drives. (Excellent cooperation is becoming established between the Institute of Machine Structures at TUB and Department of Manufacturing Technologies at the Miskolc University.)

It was in 1983 at the Gear Conference in Dresden that during the discussion following Hegyháti's lecture (Hegyháti, 1988), the question of precise grinding technology of conical worms arose. The author of this book was taking part in this debate and, partly because of his particular interest and partly at the request of the Institute of Machine Structures at TUB, he had begun intensive research in this field. As a partial result of his activity at the session of the Mechanical Committee of the Hungarian Academy convened by its Drive Technique subcommittee (Budapest, 26 May 1986), the author gave a summary (Dudás, 1986d) of his research, where he mentioned the possibility of a generally valid algorithm making possible the derivation of different types of helicoids from a common root.

2.5 SURFACE OF TOOLS

The analysis of helicoid surfaces is generally applicable to all tools having helicoidal surfaces. The aim is to create a model suitable to investigate such tools.

The side, back and complementary surfaces of tools used for machining (grinding wheels, worm milling cutters, circular module milling cutters, thread cutters, and other tools having helicoidal surfaces) are those investigated.

Books dealing with the problem of teeth generation (Erney, 1983; Litvin, 1972, Maros, Killmann and Rohonyi, 1970; Niemann and Winter, 1983; Rohonyi, 1980) generally treat the design and manufacture of tools, with correct geometry, which are indispensable to manufacturing processes, in a tangential way only.

Teeth-manufacturing tools are served very poorly in the literature – only a few publications, like Bakondi's (1974) and Drahos (1967, 1973, 1981, 1987) are available from Hungarian authors.

Currently, the role of the superhard grinding wheel and coated or hard-metal inserted tools is increasing (Bilz, 1976; Lierath and Popke, 1985; Dudás, 1987).

While helicoidal surfaces can generally be prepared using universal tools, special profile tools can only manufacture their mated elements. The geometric shape of these tools is basically determined by the helicoidal surface to be machined (Winter and Wilkesmann, 1975) so individually shaped surfaces require special tooth-generating milling cutters (Boecker and Rochel, 1965; Hermann, 1976).

The basic tool for machining helicoidal surfaces is the milling cutter or grinding wheel manufactured to the required precision ((Bluzat, 1986; Juchem, 1987; Jauch, 1960; Litvin, 1994; Niemann, 1942).

To manufacture these tools with correct geometry, the operational conditions should be mathematically investigated; that is, a thorough knowledge of machining geometry and the manufacturing process are necessary.

2.6 GENERAL CONCLUSIONS BASED ON THE LITERATURE

In general, in the publications mentioned above, the theory and practice of helicoidal surfaces are treated separately. Few theoretical experts deal with the practical problems of manufacturing and there are few practising engineers who investigate the theoretical background of practical problems. This is why problems of theory and practice common to both are handled in only a few instances.

It was in the theoretical treatment that the consideration originated, for example, of cylindrical helicoidal surfaces with constant lead thread movement of a line as a ruled surface or surfaces having been formed by generating a curve thread movement. The profile of a manufacturing tool is determined using a similar principle, repeating the process the opposite way.

Publications discuss practical problems, describing them and providing practical means for their solution. But explanations of empirically-solved problems are never evaluated.

Because theoretical and practical investigations are treated separately, not only in publications but in practice too, the results achieved in one field hardly impact on others.

This is why the author of this book strives to find correlation between theoretical discussion, deductions from kinematics and the set of tools that equip machine tools that are used to generate helicoids.

In checking the geometry of helicoidal surfaces, similar problems arise, as the relevant publications show. Helicoids, the most characteristic spatial surfaces, when checked, are generally handled as plane geometric problems (Boecker and Rochel, 1964; Klingelnberg, 1972) (for example, pitch and profile are checked in axial section or in the plane of the generator line). So the plan selected for checking on a helicoidal surface can only be based on a sampling process and the conclusion derived for the variations in the complete surface does not seem to be theoretically satisfactory (Klingelnberg, 1972).

In this book we discuss how the spatial checking of the helicoid geometry increases the reliability of assessing them to a significant measure. This stresses the need to formulate, on the basis of manufacturing geometry and technology for helicoids, a general model appropriate to the many-sided investigation problems of manufacturing technology for helicoids. This type of investigation would arrive at closely related theoretical and practical results suitable for the design, manufacture and checking of helicoids as summed up in this book.

The author has applied his results achieved in the field of manufacturing geometry both at Miskolc University and at the DIGÉP company.

Now, at the beginning of the 21st century, it can be expected that CNC-controlled teething machines and 3D-controlled coordinate-measuring devices will lead to the present tooth geometry and the techniques of tooth generation being changed drastically.

3

MANUFACTURING GEOMETRY FOR CONSTANT PITCH HELICOIDAL SURFACES

3.1 DEVELOPMENT OF MANUFACTURE OF CYLINDRICAL WORM GEAR DRIVES HAVING ARCHED PROFILE

Previous investigations (Drobni, 1968; Dudás, 1973) unambiguously showed that the technical possibilities and characteristics of worm gear drives point to the cylindrical worm with arched profile (Dudás, 1976) as capable of development and manufacture, for example in wire-drawing machines.

In this chapter the manufacturing technology of circular arch profile in axial section worm drives is summarized; the exact manufacturing details were discussed in the author's previously published works (Dudás, 1973, 1980, 1988a).

The manufacture was carried out based on a Mátrix licence using a KM-250 type machine produced by the CSEPEL Machine Tool Factory. Because of the given correction method for the grinding wheel as well as to simplify the worm checking, the circular arched profile situated originally in normal section was transerred into worm axial section (Dudás, 1973).

The possibility of transferring a circular arched profile into axial section was raised by Krivenko (1967). According to him, the worm-operating conditions would be optimal in this case. Because the problems related to precision manufacturing had not been solved, instead of making this type of worm, he manufactured a so-called 'equivalent' worm. (His personal opinion is that this type of worm can only be machined by lathe turning.)

By this way of manufacture the teething of elements is prepared with an inclined shape approximately equal to the worm described above.

The geometric dimensions in the axial section and the profile of the worm to be manufactured are given in Figure 3.1 using a $x_1y_1z_1$ constant coordinate system. The profile suggested to be manufactured as a circular arched worm profile is determined by the sizes ρ_{ax} and k in axial section, where ρ_{ax} is the radius of rounding the tooth, and K is the distance between centre line of worm and centre of circular profile.

By using a circular arched profile in an axial section worm it appears that all the equations giving the characteristic of engagement (equations of the helicoids on the worm), the description of teeth surfaces, the contact spatial curves, the generator of worm profile, and the similar equations of the worm gear teeth etc are far simpler. Tests (Krivenko, 1967) showed that the best results both for load-carrying capacity and for operational quality could be obtained

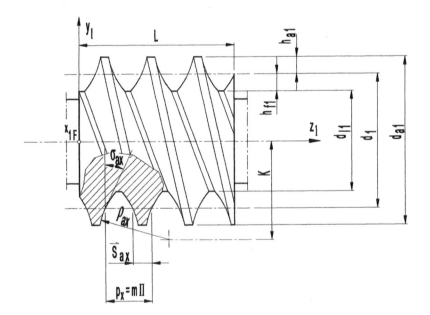

Figure 3.1 *Geometric sizes and profile of worm having circular profile in axial section*

when the generator is an ellipse or a circle. At the same time, a circular profile provides the possibility of further advantages. During manufacture, the checking both of worm and of generating milling cutter are simpler and tool profiles can be produced easier. It should be mentioned that the equations describing other types of arched profile worm gear drive are far more complicated.

3.1.1 Analysis and equation of helicoidal surface having circular profile in axial section

The cylindrical helicoidal surface is generated by a circle with radius ρ_{ax} situated in axial section. The arc is rotated round axis z_1 parallel to it; an axial displacement is realized in accordance with parameter p (see Figure 3.2).

Displacement (z_1), angular displacement ϑ and parameter p correlate to each other as:

$$z_1 = p \cdot \hat{\vartheta} \tag{3.1}$$

The parameter p is the axial displacement belonging to unit angular displacement ($\partial = 1$ rad). The points of the generating circle during one complete revolution follow different thread lines, having equal leads p_z, which is the lead of the helicoidal surface as well.

$$p_z = 2 \times \pi \times p \tag{3.2}$$

The value of parameter p is:

$$p = \frac{p_z}{2 \cdot \pi} = \frac{d_{01}}{2} \cdot tg\gamma_0 = \frac{m \cdot z_1}{2} \tag{3.3}$$

where $p_z = p_x \times z_1 = m \times p \times z_1$ – lead, γ_0 is lead angle on worm reference cylinder, z_1 is number of teeth on worm.

The helicoidal surface, having a circular generator in axial section, can be manufactured by lathe turning. The parameters required for generation and the geometry can be seen in Figure 3.2.

The face of the tool fits on plane $\eta O_3 \zeta$, its cutting edge being determined by the circle of radius ρ_{ax}. One side of the worm surface is prepared using the 1–2 edge of the moving tool. As the generator of the helicoid, the cutting edge 1–2 follows a worm path relative to the workpiece; this is why the origin (O_{1F}) of coordinate system K_{1F} (x_{1F}, y_{1F}, z_{1F}) performs a translational motion along the centre line of the workpiece within the coordinate system K_{sz} (ξ, η, ζ).

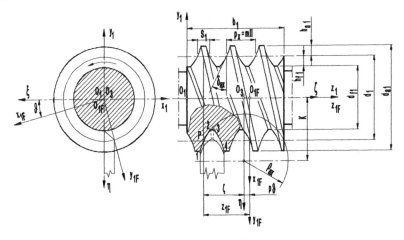

Figure 3.2 *Profile of worm having circular profile in axial section (sketch of its generation)*

The sign of parameter p can be positive or negative depending on whether a right- or left-hand worm path is followed, ie whether it is a right- or left-hand lead helicoid. Applying coordinate transformation, the correlations between systems K_{1F} (x_{1F}, y_{1F}, z_{1F}) and K_{sz} (ξ, η, ζ) based on Figures 3.2 and 3.3 can be derived.

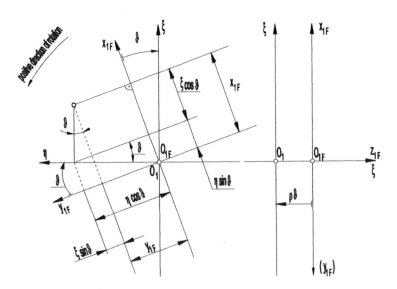

Figure 3.3 *Correlation between the coordinate systems K_{1F} (x_{1F}, y_{1F}, z_{1F}) rotating, and K_{sz} (ξ, η, ζ) joined to tool*

To investigate teeth surfaces the kinematic method is applied. For the coordinate transformation needed for use in the kinematic method, homogeneous coordinates should be used (Litvin, 1972). This is necessary because the origins of the coordinate systems are at different points. That is, the transformation consists of two parallel movements, rotational and translational (thread motion).

According to the above, the matrix of transformation from system K_{sz} (ξ, η, ζ) to system K_{1F} (x_{1F}, y_{1F}, z_{1F}) is:

$$\underline{M}_{1F,sz} = \begin{bmatrix} \cos\vartheta & \sin\vartheta & 0 & 0 \\ -\sin\vartheta & \cos\vartheta & 0 & 0 \\ 0 & 0 & 1 & -p\cdot\vartheta \\ 0 & 0 & 0 & 1 \end{bmatrix} \tag{3.4}$$

where the angular displacement (ϑ) of the workpiece during machining determines the relative position of the cutting edge η, O_3, ζ and the plane x_{1F}, y_{1F}, z_{1F} (the measure of shift).

As the generator of the profile fits the plane η, O_3, ζ (it is the axial section) the equation of the profile generator can be written

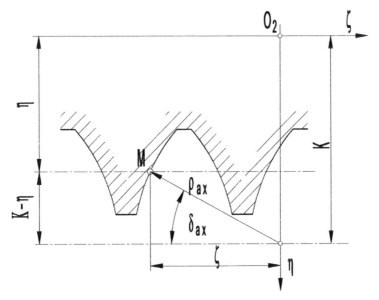

Figure 3.4 _Determination of generator in axial section_

using Figure 3.4. The coordinates of an oblique point of the profile generator can be expressed as:

$$M_j\left[0, \eta_M, -\sqrt{\rho_{ax}^2 - (K-\eta)^2}\right] \text{ on the right-hand side of the tooth}$$

$$M_b\left[0, \eta_M, +\sqrt{\rho_{ax}^2 - (K-\eta)^2}\right] \text{ on the left-hand side of the tooth}$$

(3.5)

Transformation into the coordinate system joined to the worm can be written as:

$$\vec{r}_{1F} = \underline{M}_{1F,sz} \cdot \vec{r}_{sz} \tag{3.6}$$

where $\vec{r}_{sz}$ is the equation of the tool edge or generating curve in the tool coordinate system K_{sz} (in the case of circular profile worm, it is determined by (3.5)).

After accomplishing the assigned operation, the equation of the worm surface obtained is:

$$\vec{r}_{1F} = \underline{M}_{1F,sz} \cdot \vec{r}_{sz} = \begin{bmatrix} \cos\vartheta & \sin\vartheta & 0 & 0 \\ -\sin\vartheta & \cos\vartheta & 0 & 0 \\ 0 & 0 & 1 & -p\cdot\vartheta \\ 0 & 0 & 0 & 1 \end{bmatrix} \cdot \begin{bmatrix} 0 \\ \eta \\ -\sqrt{\rho_{ax}^2 - (K-\eta)^2} \\ t_{sz} \end{bmatrix} \tag{3.7}$$

The right-hand side surface of the worm helicoid is obtained in the rotating coordinate system. Taking into consideration the direction of rotation ϑ in Figure 3.3, in this case it is negative, and the following equations are obtained:

$$\left.\begin{array}{l} x_{1F} = -\varsigma \cdot \sin\vartheta \cdot \\ y_{1F} = \varsigma \cdot \cos\vartheta \cdot \\ z_{1F} = p\cdot\vartheta - \sqrt{\tilde{n}_{ax}^2 - (K-\varsigma)^2} \\ t_{1F} = t_{sz} = 1 \end{array}\right\} \text{profile on the right-hand side}$$

$$\left.\begin{array}{l} x_{1F} = -\varsigma \cdot \sin\vartheta \cdot \\ y_{1F} = \varsigma \cdot \cos\vartheta \cdot \\ z_{1F} = p\cdot\vartheta + \sqrt{\tilde{n}_{ax}^2 - (K-\varsigma)^2} \\ t_{1F} = t_{sz} = 1 \end{array}\right\} \text{profile on the left-hand side}$$

(3.8)

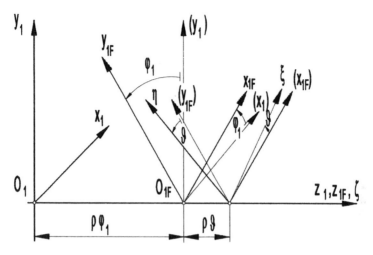

Figure 3.5 *Connection between coordinate systems, stationary K_1 and rotating K_{sz}*

From the coordinate system K_{1F}, connected to the worm, it is transferred into the stationary coordinate system K_1 using a transformation matrix according to Figure 3.5:

$$\underline{M}_{1,1F} = \begin{bmatrix} \cos\varphi_1 & -\sin\varphi_1 & 0 & 0 \\ \sin\varphi_1 & \cos\varphi_1 & 0 & 0 \\ 0 & 0 & 1 & p\cdot\varphi_1 \\ 0 & 0 & 0 & 1 \end{bmatrix} \qquad (3.9)$$

The transformation is:

$$\vec{r}_1 = \underline{M}_{1,1F} \cdot \vec{r}_{1F} \qquad (3.10)$$

After substituting it is:

$$\vec{r}_1 = \begin{bmatrix} \cos\varphi_1 & -\sin\varphi_1 & 0 & 0 \\ \sin\varphi_1 & \cos\varphi_1 & 0 & 0 \\ 0 & 0 & 1 & p\cdot\varphi_1 \\ 0 & 0 & 0 & 1 \end{bmatrix} \cdot \begin{bmatrix} -\eta\cdot\sin\vartheta \\ \eta\cdot\cos\vartheta \\ p\cdot\vartheta - \sqrt{\rho_{ax}^2 - (K-\eta)^2} \\ 1 \end{bmatrix} \qquad (3.11)$$

where φ_1 denotes the value of angular displacement between the co-ordinate systems K_{1F} and K_1, and $(p \times \varphi_1) = z_1$ is the axial translation of helicoid surface along the axis owing to angular displacement φ_1.

Having accomplished the operations, the equation system of the right-side surface of the worm in the stationary coordinate system is:

$$
\left.
\begin{aligned}
x_1 &= -\eta \cdot \sin \vartheta \cdot \cos \varphi_1 - \eta \cdot \cos \vartheta \cdot \sin \varphi_1 = -\eta \cdot \sin(\vartheta + \varphi_1) \\[2mm]
y_1 &= \eta \cdot \cos \vartheta \cdot \cos \varphi_1 - \eta \cdot \sin \vartheta \cdot \sin \varphi_1 = \eta \cdot \cos(\vartheta + \varphi_1) \\[2mm]
z_1 &= p \cdot (\vartheta + \varphi_1) - \sqrt{\rho_{ax}^2 - (K - \eta)^2}
\end{aligned}
\right\}
\tag{3.12}
$$

When shifting the stationary coordinate system on the worm body so that plane $x_1 O_1 y_1$ is the symmetry plane of the tooth, the changing value of z_{ax} should be added to the value of the z coordinate modifying equation (3.12).

Introducing the symbol $\vartheta + \varphi_1 = \Theta$ (see Figure 3.6)

$$
\left.
\begin{aligned}
x_1 &= -\eta \cdot \sin \Theta \\
y_1 &= \eta \cdot \cos \Theta \\
z_1 &= p \cdot \Theta - \sqrt{\rho_{ax}^2 - (K - \eta)^2} + z_{ax}
\end{aligned}
\right\}
\tag{3.13}
$$

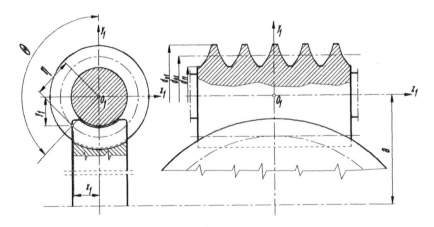

Figure 3.6 *Contact area of worm and worm gear, correlation between parameters η and θ*

It is sufficient to investigate the helicoidal surface of the worm on the worm gear side, situated below plane $x_1 O_1 z_1$ and within the contact area (see Figure 3.6).

The locus of points on this part of the worm should fulfil the inequalities:

$$90° \le \theta \le 270°$$
$$r_{fl} \le \eta \le r_{al}$$

where r_{al} is the radius of the worm addendum cylinder and r_{fl} is the radius of the worm dedendum cylinder.

The value of z_{ax} can be obtained by substituting $\theta = 180°$, $z_1 = \overline{S_{ax}}/2$ and the value ζ from Figure 3.4 into the third equation of the equation system (3.13):

$$\zeta = \sqrt{\rho_{ax}^2 - (K - \eta)^2} = \rho_{ax} \cos \delta_{ax} \tag{3.14}$$

$$\left. \begin{array}{l} z_{ax} = -\dfrac{S_{ax}}{2} - p \cdot \pi - \rho_{ax} \cdot \cos \delta_{ax} \quad \text{right-hand profile} \\[3mm] z_{ax} = \dfrac{S_{ax}}{2} - p \cdot \pi + \rho_{ax} \cdot \cos \delta_{ax} \quad \text{left-hand profile} \end{array} \right\} \tag{3.15}$$

where S_{ax} is the tooth thickness of the reference cylinder in its axial section, and δ_{ax} is the profile angle of the reference cylinder in its axial section.

The right- and left-hand profiles of the thread are determined by equations (3.13) and (3.14) in the stationary coordinate system K_1 (x_1, y_1, z_1). The curve of the left-hand profile is situated symmetrically with respect to axis $O_1 y_1$. Therefore, by rolling it over by $180°$ (between supports), the other side can be machined within the same kinematic conditions.

The equations above can be used for calculation of tool profile (eg milling cutter etc), for manufacturing go and no go gauges, and for investigation of conditions of tooth-mating processes etc. The use of a circular profile in the axial section of the worm, shown in Figure 3.2, is justified: the equations determine contact area (equations of helicoids on worm), equations of helicoids of worm gear, contact lines, profile generator for the worm. The corresponding equations referring to the teeth of the worm gear as well as the equation expressing tooth thickness are simpler, making for easier calculations of geometric sizes and strength dimensioning. Profile gauges necessary for machining of the worm, and suitable for the checking of milling cutters, and profiles of other types of worm-generating tools can be determined, using simple equations. Also, they are easy to machine. It should be mentioned that previously derived equations valid for other types of arched profile threads are far more complicated and extensive.

3.1.2 Analysis of worm manufacturing finishing; an exact solution

An exact helicoidal surface for worms can be created by lathe cutting of the threads. When the thread surface is ground using a disc-like tool, especially with high lead worms on a ready-made helicoidal surface, because of the nature of the grinding wheel, new notches are created. The worms manufactured by this method cannot possess true geometry.

The aim is to determine the profile of the grinding wheel so that the profile of the worm surface in its axial section is always the required and prescribed one. To attain this, the edges of the grinding wheel, which remove the missing parts from the whole thread profile, should be cleared away. If a diamond profile worm was made available, similar to the one to be manufactured, profiling of the grinding wheel would be simple. This hypothetical tool would remove the parts creating the missing profile.

As of now, this tool does not exist and this diamond worm would be too expensive to make and, therefore, not suitable for production purposes. A grinding wheel profile controller (generator) device designed by the author could, it is suggested, replace that hypothetical method mentioned before. The essence of the device is that the generator circle radius (ρ_{ax}) of the worm to be ground is adjustable in different, given positions and rolled down before the grinding wheel.

Let the grinding wheel axes be inclined by lead angle γ_o, and the device controlling the wheel (generating device) be situated in line with the worm (see Figures 3.7 and 3.8).

To avoid undercutting, especially at high leads, it is likely that the condition $\gamma > \gamma_o$ deviation should be applied.

The distance between axes of the grinding wheels I–I and of worm III–III using the known data (see Figure 3.8) can be calculated as in equation (2.2) as:

$$A_{sz} = K + r_{sz} - h_{sz} \qquad (3.16)$$

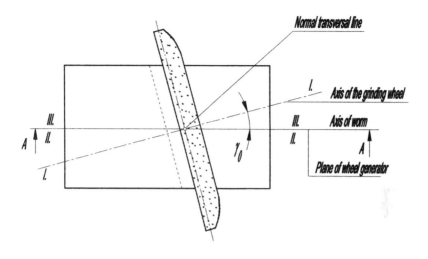

Figure 3.7 _Inclination of grinding wheel with_ γ_o _lead angle on reference cylinder (see Dudás, 1976)_

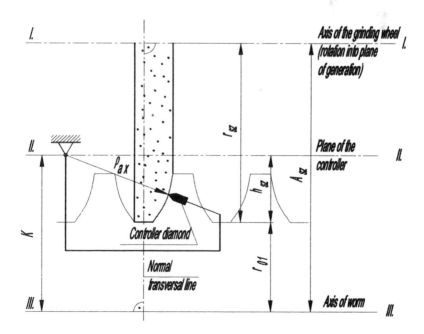

Figure 3.8 _Positioning of wheel profile controller in the main plane (section A–A in Figure 3.7) (as suggested by Dudás, 1976)_

3.1.3 Problems of manufacturing geometry during final machining of worm – determination of grinding wheel profile

3.1.3.1 Equation system describing surface of grinding wheel

The shaping of a helicoidal surface using a grinding wheel can be regarded as the equivalent of a worm and worm wheel drive when the crossing angle between axes of elements is equal to the angle of inclination of the grinding wheel (γ_o).

To determine the equation of a grinding wheel as an enveloping surface under the conditions mentioned above, the side surface of the transfer load should be in continuous contact during the operation. As these surfaces are described by an independent parameter, a method to determine the wrapping surface of the series of a single parameter surface should be investigated. As with the investigation of the tooth surface, homogeneous coordinates have to be used for the necessary coordinate transformations.

The two parameters (η, ϑ) (equation (3.7)) of the worm surface are known in the rotating coordinate system K_{1F} (x_{1F}, y_{1F}, z_{1F}) as well as the relative displacement for worm and grinding wheel in the stationary coordinate system K_1 (x_1, y_1, z_1).

The worm surface in rotating coordinate system K_{1F} (x_{1F}, y_{1F}, z_{1F}) should be transferred into the grinding wheel coordinate system K_{2F} (x_{2F}, y_{2F}, z_{2F}) to obtain equations for the series of worm surfaces in terms of a parameter φ_1, the angular displacement coefficient.

Then the equation expressing conditions for the contact of teeth surfaces (worm to grinding wheel) should be determined.

The equation expressing contact conditions and equations for the series of surfaces have a common solution, in other words, the equations of the wrapping wheel. The tooth profile for element No2 should be determined as the wrapping curve of the tooth profile for element No1, developed by their relative motions.

Rotating the characteristics around axis z_2, the surface of the grinding wheel is obtained. Knowing that the surface of element No1 (3.7) is the worm, it can be determined which surface should be shaped on element No2 by applying motion projection so that surfaces A_1 and A_2 are in continuous line contact so as to transfer the motion following the rule:

$$i_{21} = \frac{d\varphi_1}{d\varphi_2} = \frac{\omega^{(1)}}{\omega^{(2)}} = \text{constant} \tag{3.17}$$

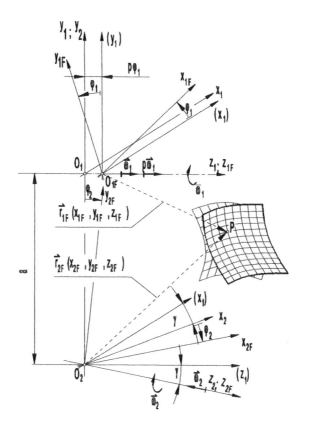

Figure 3.9 *Connection between coordinate systems to investigate process grinding*

To investigate drive between skew axes, at least three coordinate systems are needed (Figure 3.9). In this case, four coordinate systems are used.

The coordinate systems applied are (they all are right direction systems):

- K_{1F} – coordinate system fix connected to worm body. The two-parameter (η, ϑ) equation of the helicoid is formulated here;
- K_1 – stationary coordinate system connected with stand with centre line z_1 coinciding with axis of worm body z_{1F};
- K_2 – stationary coordinate system connected with stand with centre line z_2 coinciding with axis of wheel z_{2F};
- K_{2F} – coordinate system connected to grinding wheel with centre line z_{2F} inclining to axis of worm body z_{1F} with angle γ_o.

The origin of the coordinate system joined to element No1, ie when $t = 0$, coincides with the origin of the stationary coordinate system K_1. The stationary coordinate system K_2 joined to element No2 is placed by comparison with the stationary system K_1, the vector being defined by:

$$\vec{r}_{12} = \overrightarrow{O_1O_2} = \begin{bmatrix} 0 \\ -a \\ 0 \end{bmatrix} = \text{constant} \tag{3.18}$$

These axes of elements are z_1 and z_2; the elements rotate round them with angular speeds:

$$\omega_1 = \omega_z^{(1)} = \omega^{(1)} = 1 \left[s^{-1} \right] \tag{3.19}$$

$$\omega_2 = \omega_z^{(2)} = \omega^{(2)} = i_{21} \cdot \omega_1 \left[s^{-1} \right] \tag{3.20}$$

and at $t = t_o = 0$ they occupy angular positions $\varphi_1 = 0$ and $\varphi_2 = 0$.

As the wrapped worm surface No1 is described by:

$$\vec{r}_{1F} = \vec{r}_{1F}(\eta; \vartheta) \tag{3.21}$$

the two-parameter vector–scalar function in the coordinate system K_{1F}, introducing homogeneous coordinates of the position vectors at the same point in the different coordinate systems, can be written, using suitable transformation matrices, as:

$$\vec{r}_1 = \underline{M}_{1,1F} \cdot \vec{r}_{1F} \tag{3.22}$$

$$\vec{r}_2 = \underline{M}_{2,1F} \cdot \vec{r}_{1F} \tag{3.23}$$

$$\vec{r}_{2F} = \underline{M}_{2F,1F} \cdot \vec{r}_{1F} \tag{3.24}$$

where the transformation matrices are:

- $\underline{M}_{1,1F}$, the transformation matrix between rotating coordinate system K_{1F} and stationary system K_1;
- $\underline{M}_{2,1F} = \underline{M}_{2,1} \times \underline{M}_{1,1F}$, the transformation matrix between rotating coordinate system K_{1F} and stationary system K_2 joined to the grinding wheel;
- $\underline{M}_{2F,1F} = \underline{M}_{2F,2} \times \underline{M}_{2,1} \times \underline{M}_{1,1F}$, the transformation matrix between rotating systems K_{1F} and K_{2F};

■ $\bar{r}_1, \bar{r}_2, \bar{r}_3$, position vectors of the same point in coordinate systems K_1, K_2 and K_3.

Transformation matrices, as they denote symbolic connections, are identified as:

$$\underline{M}_{1,IF} : K_{IF} \rightarrow K_1 \qquad \underline{M}_{IF,1} : K_1 \rightarrow K_{IF}$$
$$\underline{M}_{2,1} : K_1 \rightarrow K_2 \qquad \underline{M}_{1,2} : K_2 \rightarrow K_1$$
$$\underline{M}_{2,2F} : K_{2F} \rightarrow K_2 \qquad \underline{M}_{2F,2} : K_2 \rightarrow K_{2F}$$

The connection between coordinate systems $K_{IF} \rightarrow K_1$ is shown in Figure 3.10.

$$\underline{M}_{1,IF} = \begin{bmatrix} \cos\varphi_1 & -\sin\varphi_1 & 0 & 0 \\ \sin\varphi_1 & \cos\varphi_1 & 0 & 0 \\ 0 & 0 & 1 & p\cdot\varphi_1 \\ 0 & 0 & 0 & 1 \end{bmatrix} \qquad (3.25)$$

$$\underline{M}_{1F,1} = \begin{bmatrix} \cos\varphi_1 & \sin\varphi_1 & 0 & 0 \\ -\sin\varphi_1 & \cos\varphi_1 & 0 & 0 \\ 0 & 0 & 1 & -p\cdot\varphi_1 \\ 0 & 0 & 0 & 1 \end{bmatrix} \qquad (3.26)$$

Connection between coordinate systems K_1 and K_2 is determined by Figure 3.11.

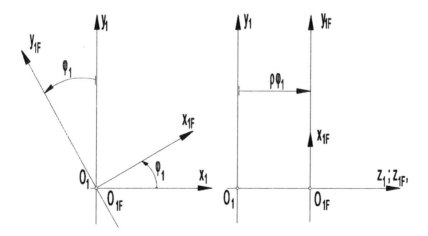

Figure 3.10 _Connection between coordinate systems K_{IF} and K_1_

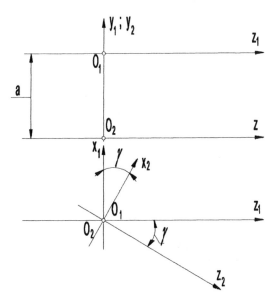

Figure 3.11 *Connection between coordinate systems K_1 and K_2*

For the purposes of simplification, the mathematics formula let γ be used instead of γ_o ($\gamma_o = \gamma$):

$$\underline{M}_{2,1} = \begin{bmatrix} \cos\gamma & 0 & \sin\gamma & 0 \\ 0 & 1 & 0 & a \\ -\sin\gamma & 0 & \cos\gamma & 0 \\ 0 & 0 & 0 & 1 \end{bmatrix} \qquad (3.27)$$

$$\underline{M}_{1,2} = \begin{bmatrix} \cos\gamma & 0 & -\sin\gamma & 0 \\ 0 & 1 & 0 & -a \\ \sin\gamma & 0 & \cos\gamma & 0 \\ 0 & 0 & 0 & 1 \end{bmatrix} \qquad (3.28)$$

The connection between coordinate systems K_2 and K_{2F} can be seen in Figure 3.12.

$$\underline{M}_{2F,2} = \begin{bmatrix} \cos\varphi_2 & -\sin\varphi_2 & 0 & 0 \\ \sin\varphi_2 & \cos\varphi_2 & 0 & 0 \\ 0 & 0 & 1 & 0 \\ 0 & 0 & 0 & 1 \end{bmatrix} \qquad (3.29)$$

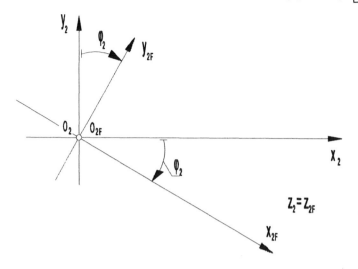

Figure 3.12 _Connection between coordinate systems K_2 and K_{2F}_

$$\underline{\mathbf{M}}_{2,2F} = \begin{bmatrix} \cos\varphi_2 & \sin\varphi_2 & 0 & 0 \\ -\sin\varphi_2 & \cos\varphi_2 & 0 & 0 \\ 0 & 0 & 1 & 0 \\ 0 & 0 & 0 & 1 \end{bmatrix}$$ (3.30)

To derive equation (3.24), matrix $\underline{\mathbf{M}}_{2F,1F}$ and matrix $\underline{\mathbf{M}}_{2F,2}$ will be used:

$$\underline{\mathbf{M}}_{2F,1F} = \underline{\mathbf{M}}_{2F,2} \times \underline{\mathbf{M}}_{2,1} \times \underline{\mathbf{M}}_{1,1F} = \underline{\mathbf{M}}_{2F,2} \times \underline{\mathbf{M}}_{2,1F}$$ (3.31)

Substituting known matrices we obtain:

$$\underline{\mathbf{M}}_{2F,1F} =$$
$$= \begin{bmatrix} \cos\varphi_2 & -\sin\varphi_2 & 0 & 0 \\ \sin\varphi_2 & \cos\varphi_2 & 0 & 0 \\ 0 & 0 & 1 & 0 \\ 0 & 0 & 0 & 1 \end{bmatrix} \cdot \begin{bmatrix} \cos\gamma\cdot\cos\varphi_1 & -\cos\gamma\cdot\sin\varphi_1 & \sin\gamma & p\cdot\varphi_1\cdot\sin\gamma \\ \sin\varphi_1 & \cos\varphi_1 & 0 & a \\ -\sin\gamma & \varphi_1 & \sin\gamma & \varphi_1 & \cos\gamma & p\cdot\varphi_1\cdot\cos\gamma \\ 0 & 0 & 0 & 1 \end{bmatrix}$$ (3.32)

Accomplishing the assigned operations for $\underline{\mathbf{M}}_{2F,1F}$, equation (3.33) is obtained:

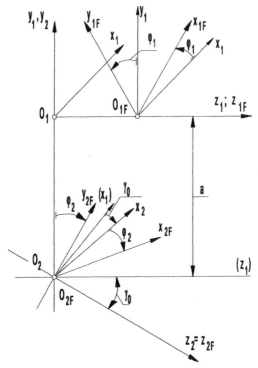

Figure 3.13 *Connection between coordinate systems K_{1F} and K_{2F}*

$$
\underline{M}_{2F,1F} =
\begin{bmatrix}
\cos\gamma\cdot\cos\varphi_1\cdot\cos\varphi_2 - & -\cos\gamma\cdot\sin\varphi_1\cdot\cos\varphi_2 - & \sin\gamma\cdot\cos\varphi_2 & p\varphi_1\cdot\sin\gamma\cdot\cos\phi_2 - \\
-\sin\varphi_1\cdot\sin\varphi_2 & -\cos\varphi_1\cdot\sin\varphi_2 & & -a\cdot\sin\varphi_2 \\
 & & & \\
\cos\gamma\cdot\cos\varphi_1\cdot\cos\varphi_2 + & -\cos\gamma\cdot\sin\varphi_1\cdot\cos\varphi_2 + & \sin\varphi_2\cdot\sin\gamma & p\varphi_1\cdot\sin\gamma\cdot\sin\phi_2 + \\
+\sin\varphi_1\cdot\sin\varphi_2 & +\cos\varphi_1\cdot\sin\varphi_2 & & +a\cdot\cos\varphi_2 \\
 & & & \\
-\sin\gamma\cdot\cos\phi_1 & \sin\gamma\cdot\sin\varphi_1 & \cos\gamma & p\varphi_1\cdot\cos\gamma \\
0 & 0 & 0 & 1
\end{bmatrix}
$$

$$(3.33)$$

The two-parameter vector-scalar function $\vec{r}_{1F} = \vec{r}_{1F}$ (η, ϑ) describing the given helicoidal surface can be transferred from system K_{1F} into system K_{2F}, using already known elements:

$$r_{2F} = \underline{M}_{2F,1F} \cdot r_{1F} \Bigg\}$$
$$\widehat{\varphi}_2 = i_{21} \cdot \widehat{\varphi}_1 \qquad (3.34)$$

On teeth surfaces of mated elements, the common contact curve of mutually enveloping surfaces fulfilling the first law of connection can be written:

$$\vec{n}_{1F} \cdot \vec{v}_{1F}^{(12)} = \vec{n}_{2F} \cdot \vec{v}_{2F}^{(12)} = \vec{n} \cdot \vec{v}^{(12)} = 0 \qquad (3.35)$$

which can be determined by the common solution of equations of contact and vector–scalar function determining teeth surfaces (3.21). To formulate the equation of contact, first knowledge of relative velocities between elements is required.

The grinding worm associated with the grinding wheel realizes a thread motion. The equation of the thread surface taking into account the boundary conditions at starting is:

$$\vec{r}_{1F} = \vec{r}_{1F}(\eta, \vartheta) \qquad (3.36)$$

To determine the relative velocity of system K_{1F} to system K_2, first by differentiating function (3.34) with respect to time, the relative velocity of system K_{1F} to system K_{2F} is:

$$\dot{\vec{r}}_{2F} = \vec{v}_{2F}^{(12)} = \frac{d}{dt}\left(\underline{M}_{2F,1F} \cdot \vec{r}_{1F}\right) = \frac{d\underline{M}_{2F,1F}}{dt} \cdot \vec{r}_{1F} \qquad (3.37)$$

The position vector $\vec{r}_{1F}$, according to equation (3.7), is time- independent. Taking into consideration the correlation between velocity vectors and expressing relative motion between systems K_{1F} and K_{2F}:

$$v_{1F}^{(12)} = \underline{M}_{1F,2F} \cdot v_{2F}^{(12)} \qquad (3.38)$$

the relative velocity vector in system K_{1F} can be obtained:

$$\vec{v}_{1F}^{(12)} = \underline{M}_{1F,2F} \cdot \frac{d\underline{M}_{2F,1F}}{dt} \cdot \vec{r}_{1F} \qquad (3.39)$$

Differentiating elements of transformation matrix $\underline{M}_{2F,1F}$ (3.33) with respect to time and introducing $\varphi_1 = (\omega_1^{(1)})t$, and choosing $\omega_1^{(1)} = 1$ for $\varphi_1 = t$ and $\varphi_2 = i_{21}t$, (3.42) is obtained.

This matrix (3.42) is transformed into the K_{1F} system using equation (3.39). The result is the product of matrices, calculated as:

$$\underline{P_1} = \underline{M_{1F,2F}} \cdot \frac{d}{dt} \underline{M_{2F,1F}} \tag{3.40}$$

The P_1 matrix is the 'kinematic transformer' matrix. First, the product of matrices has to be determined, ie:

$$\underline{M_{1F,2F}} = \underline{M_{1F,1}} \underline{M_{1,2}} \underline{M_{2,2F}} \tag{3.41}$$

Knowing matrix P_1, the relative velocity vector can be calculated using (3.39). After substitution:

$$v_{1Fx}^{(12)} = -\eta \cdot \cos \vartheta \cdot (1 + i_{21} \cdot \cos \gamma) +$$
$$+ i_{21} \cdot \sin \gamma \cdot \sin \varphi_1 \cdot \left(p \cdot \vartheta - \sqrt{\rho_{ax}^2 - (K - \eta)^2} + c + p \cdot \varphi_1 \right) - i_{21} \cdot a \cdot \cos \gamma \cdot \cos \varphi_1 \tag{3.46}$$

$$v_{1Fy}^{(12)} = -\eta \cdot \sin \vartheta \cdot (1 + i_{21} \cdot \cos \gamma) +$$
$$+ i_{21} \cdot \sin \gamma \cdot \cos \varphi_1 \cdot \left(p \cdot \vartheta - \sqrt{\rho_{ax}^2 - (K - \eta)^2} + c + p \cdot \varphi_1 \right) - i_{21} \cdot a \cdot \cos \gamma \cdot \sin \varphi_1 \tag{3.47}$$

$$v_{1Fz}^{(12)} = -\eta \cdot i_{21} \cdot \sin \gamma \cdot \cos (\vartheta + \varphi_1) - i_{21} \cdot a \cdot \sin \gamma + p \tag{3.48}$$

At the contact point of the mutual enveloping teeth profiles, the vector of relative velocity should be perpendicular to the common normal vector.

Knowing the relative velocity conditions between worm and grinding wheel, the equation of contact can be written:

$$\vec{n}_{1F} \cdot \vec{v}_{1F}^{(12)} = 0 \tag{3.49}$$

where $\overline{n}_{1F}$ is the normal vector of surface A_1.

$$\vec{n}_{1F} = \frac{\partial \vec{r}_{1F}}{\partial \eta} \times \frac{\partial \vec{r}_{1F}}{\partial \vartheta} = \begin{vmatrix} \vec{i} & \vec{j} & \vec{k} \\ \dfrac{\partial x_{1F}}{\partial \eta} & \dfrac{\partial y_{1F}}{\partial \eta} & \dfrac{\partial z_{1F}}{\partial \eta} \\ \dfrac{\partial x_{1F}}{\partial \vartheta} & \dfrac{\partial y_{1F}}{\partial \vartheta} & \dfrac{\partial z_{1F}}{\partial \vartheta} \end{vmatrix} \tag{3.50}$$

$$\frac{d}{dt}\underline{M}_{2F,1F} =$$

$$
\begin{bmatrix}
\begin{array}{l}-\sin\varphi_1\cdot\cos\varphi_2\cdot(i_{21}+\cos\gamma)-\\ \cos\varphi_1\sin\varphi_2\cdot(1+i_{21}\cdot\cos\gamma)\end{array} & \begin{array}{l}-\cos\varphi_1\cdot\cos\varphi_2\cdot(i_{21}+\cos\gamma)+\\ \sin\varphi_1\sin\varphi_2\cdot(1+i_{21}\cdot\cos\gamma)\end{array} & -i_{21}\cdot\sin\gamma\cdot\sin\varphi_2 & \begin{array}{l}p\cdot\sin\gamma\cdot\cos\varphi_2-\\ i_{21}\cdot p\cdot\varphi_1\cdot\sin\gamma\cdot\sin\varphi_2-\\ i_{21}\cdot a\cdot\sin\varphi_2\end{array}\\[2em]
\begin{array}{l}-\sin\varphi_1\cdot\cos\varphi_2\cdot(i_{21}+\cos\gamma)+\\ \cos\varphi_1\sin\varphi_2\cdot(1+i_{21}\cdot\cos\gamma)\end{array} & \begin{array}{l}-\cos\varphi_1\cdot\sin\varphi_2\cdot(i_{21}+\cos\gamma)-\\ \sin\varphi_1\cos\varphi_2\cdot(1+i_{21}\cos\gamma)\end{array} & i_{21}\cdot\sin\gamma\cdot\cos\varphi_2 & \begin{array}{l}p\cdot\sin\gamma\cdot\sin\varphi_2+\\ i_{21}\cdot p\cdot\varphi_1\cdot\sin\gamma\cdot\cos\varphi_2-\\ i_{21}\cdot a\cdot\sin\varphi_2\end{array}\\[2em]
\sin\gamma\cdot\sin\varphi_1 & \sin\gamma\cdot\cos\varphi_1 & 0 & p\cdot\cos\gamma\\[1em]
0 & 0 & 0 & 0
\end{bmatrix}
$$

$$(3.42)$$

$$
\underline{M}_{1F,2F} =
\begin{bmatrix}
\cos\gamma\cdot\cos\varphi_1\cdot\cos\varphi_2 - \sin\varphi_1\cdot\sin\varphi_2 & \cos\gamma\cdot\cos\varphi_1\cdot\sin\varphi_2 + \sin\varphi_1\cdot\cos\varphi_2 & -\cos\varphi_1\cdot\sin\gamma & -a\cdot\sin\varphi_1 \\
-\cos\gamma\cdot\sin\varphi_1\cdot\cos\varphi_2 - \cos\varphi_1\cdot\sin\varphi_2 & -\cos\gamma\cdot\sin\varphi_1\cdot\sin\varphi_2 + \cos\varphi_1\cdot\cos\varphi_2 & \sin\varphi_1\cdot\sin\gamma & -a\cdot\cos\varphi_1 \\
\sin\gamma\cdot\cos\varphi_2 & \sin\gamma\cdot\sin\varphi_2 & \cos\gamma & -p\cdot\varphi_1 \\
0 & 0 & 0 & 1
\end{bmatrix}
$$

$$(3.43)$$

$$\frac{d}{dt}\underline{M}_{1F,2F} =$$

$$
\begin{bmatrix}
\begin{aligned}
&-\sin\varphi_1\cdot\cos\varphi_2\cdot(\cos\gamma+i\cdot\cos\alpha)-\\
&-\cos\varphi_1\cdot\sin\varphi_2\cdot(i\cdot\cos\gamma+\cos\alpha)-\\
&-\sin\varphi_1\cdot\sin\varphi_2\cdot\sin\gamma\cdot\sin\alpha+\\
&+i\cdot\cos\varphi_1\cdot\cos\varphi_2\cdot\sin\gamma\cdot\sin\alpha;
\end{aligned}
&
\begin{aligned}
&-\sin\varphi_1\cdot\sin\varphi_2\cdot(\cos\gamma+i\cdot\cos\alpha)+\\
&+\cos\varphi_1\cdot\cos\varphi_2\cdot(\cos\alpha+i\cdot\cos\gamma)-\\
&-\cos\varphi_2\cdot\sin\varphi_1\cdot\sin\gamma\cdot\sin\alpha+\\
&+i\cdot\sin\varphi_2\cdot\cos\varphi_1\cdot\sin\gamma\cdot\sin\alpha;
\end{aligned}
&
\begin{aligned}
&+\sin\varphi_1\cdot\sin\gamma\cdot\cos\alpha-\\
&-\cos\varphi_1\cdot\sin\alpha;
\end{aligned}
&
\begin{aligned}
&c\cdot\sin\varphi_1-\\
&-a\cdot\cos\varphi_1
\end{aligned}
\\[2em]
\begin{aligned}
&-\cos\varphi_1\cdot\sin\varphi_2\cdot(\cos\gamma+i\cdot\cos\alpha)+\\
&+\sin\varphi_1\cdot\sin\varphi_2\cdot(i\cdot\cos\gamma+\cos\alpha)-\\
&-\sin\varphi_2\cdot\cos\varphi_1\cdot\sin\gamma\cdot\sin\alpha-\\
&-i\cdot\cos\varphi_2\cdot\sin\varphi_1\cdot\sin\gamma\cdot\sin\alpha;
\end{aligned}
&
\begin{aligned}
&-\cos\varphi_1\cdot\sin\varphi_2\cdot(\cos\gamma+i\cdot\cos\alpha)-\\
&-\sin\varphi_1\cdot\cos\varphi_2\cdot(\cos\alpha+i\cdot\cos\gamma)+\\
&+\cos\varphi_1\cdot\cos\varphi_2\cdot\sin\gamma\cdot\sin\alpha+\\
&+i\cdot\sin\varphi_1\cdot\sin\varphi_1\cdot\sin\gamma\cdot\sin\alpha;
\end{aligned}
&
\begin{aligned}
&+\cos\varphi_1\cdot\sin\gamma\cdot\cos\alpha+\\
&+\sin\varphi_1\cdot\sin\alpha;
\end{aligned}
&
\begin{aligned}
&-c\cdot\cos\varphi_1+\\
&+a\cdot\sin\varphi_1
\end{aligned}
\\[2em]
\begin{aligned}
&-i\cdot\sin\gamma\cdot\sin\varphi_2-\\
&-i\cdot\cos\varphi_2\cdot\cos\gamma\cdot\sin\alpha;
\end{aligned}
&
\begin{aligned}
&-i\cdot\sin\gamma\cdot\cos\varphi_2-\\
&-i\cdot\sin\varphi_2\cdot\cos\gamma\cdot\sin\alpha;
\end{aligned}
& 0 & -p \\[1em]
0 & 0 & 0 & 1
\end{bmatrix}
$$

$$(3.44)$$

$$\underline{P_1} = \begin{bmatrix}
1+i_{21}\cdot\cos\gamma & 0 & i_{21}\cdot\sin\gamma\cdot\sin\varphi_1 & -i_{21}\cdot a\cdot\cos\gamma\cdot\cos\varphi_1 + i_{21}\cdot p\cdot\varphi_1\cdot\sin\gamma\cdot\sin\varphi_1 \\
0 & -(1+i_{21}\cdot\cos\gamma) & i_{21}\cdot\sin\gamma\cdot\cos\varphi_1 & i_{21}\cdot a\cdot\cos\gamma\cos\varphi_1 + i_{21}\cdot p\cdot\varphi_1\cdot\sin\gamma\cdot\cos\varphi_1 \\
-i_{21}\cdot\sin\gamma\cdot\sin\varphi_1 & -i_{21}\cdot\sin\gamma\cdot\cos\varphi_1 & 0 & -i_{21}\cdot a\cdot\sin\gamma_1 + p \\
0 & 0 & 0 & 0
\end{bmatrix}$$

$$(3.45)$$

where the position vector $\overline{r}_{1F}$ scalar components are x_{1F}, y_{1F}, z_{1F}.

The normal vector $\vec{n}_{1F}$ can be calculated by development of determinant (3.50):

$$\left.\begin{array}{l} n_{1Fx} = \dfrac{\partial y_{1F}}{\partial \eta} \cdot \dfrac{\partial z_{1F}}{\partial \vartheta} - \dfrac{\partial y_{1F}}{\partial \vartheta} \cdot \dfrac{\partial z_{1F}}{\partial \eta} \\[2mm] n_{1Fy} = -\left[\dfrac{\partial x_{1F}}{\partial \eta} \cdot \dfrac{\partial z_{1F}}{\partial \vartheta} - \dfrac{\partial x_{1F}}{\partial \vartheta} \cdot \dfrac{\partial z_{1F}}{\partial \eta} \right] \\[2mm] n_{1Fz} = \dfrac{\partial x_{1F}}{\partial \eta} \cdot \dfrac{\partial y_{1F}}{\partial \vartheta} - \dfrac{\partial x_{1F}}{\partial \vartheta} \cdot \dfrac{\partial y_{1F}}{\partial \eta} \end{array}\right\} \qquad (3.51)$$

Determining coordinates of normal vector $\vec{n}_{1F}$ and substituting into equation (3.51), the result is:

$$\frac{\partial x_{1F}}{\partial \eta} = -\sin \vartheta \qquad (3.52)$$

$$\frac{\partial y_{1F}}{\partial \eta} = \cos \vartheta \qquad (3.53)$$

$$\frac{\partial z_{1F}}{\partial \eta} = -\frac{K - \eta}{\sqrt{\rho_{ax}^2 - (K - \eta)^2}} \qquad (3.54)$$

$$\frac{\partial x_{1F}}{\partial \vartheta} = -\eta \cdot \cos \vartheta \qquad (3.55)$$

$$\frac{\partial y_{1F}}{\partial \vartheta} = -\eta \cdot \sin \vartheta \qquad (3.56)$$

$$\frac{\partial z_{1F}}{\partial \vartheta} = p \qquad (3.57)$$

After substituting :

$$\left.\begin{array}{l} \vec{n}_{1Fx} = -\eta \cdot \sin \vartheta \cdot \dfrac{K - \eta}{\sqrt{\rho_{ax}^2 - (K - \eta}^2} + p \cdot \cos \vartheta \\[3mm] \vec{n}_{1Fy} = \eta \cdot \cos \vartheta \cdot \dfrac{K - \eta}{\sqrt{\rho_{ax}^2 - (K - \eta}^2} + p \cdot \sin \vartheta \\[3mm] \vec{n}_{1Fz} = \eta \end{array}\right\} \qquad (3.58)$$

Substituting equations (3.58), (3.46), (3.47) and (3.48) into equation (3.49), the condition of contact can be expressed as:

$$\vec{n}_{1F} \cdot \vec{v}_{1F}^{(12)} = 0 = \eta \cdot \frac{K - \eta}{\sqrt{\rho_{ax}^2 - (K - \eta)^2}} \cdot \sin \gamma \cdot \cos (\vartheta + \varphi_1) \cdot z_1 + \left.\begin{array}{l} \\ + \eta \cdot a \cdot \dfrac{K - \eta}{\sqrt{\rho_{ax}^2 - (K - \eta)^2}} \cdot \cos \gamma \cdot \sin (\vartheta + \varphi_1) - \\[2mm] - \eta \cdot p \cdot \cos \gamma + p \cdot \sin \gamma \cdot \sin (\vartheta + \varphi_1) \cdot z_1 - \\[1mm] - p \cdot a \cdot \cos \gamma \cdot \cos (\vartheta + \varphi_1) - \\[1mm] - \eta^2 \cdot \sin \gamma \cdot \cos (\vartheta + \varphi_1) - \eta \cdot a \cdot \sin \gamma \end{array}\right\} \quad (3.59)$$

Equation (3.59) determines the correlation between parameters (η, ϑ) of the contact point for worm and grinding wheel and the value of angular displacement φ_1.

To determine the contact curve in the system K_{1F} the following equations should be used:

$$\left.\begin{array}{l} \vec{n}_{1F} \cdot \vec{v}_{1F}^{(12)} = 0 \\ x_{1F} = x_{1F} (\eta, \vartheta) \\ y_{1F} = y_{1F} (\eta, \vartheta) \\ z_{1F} = z_{1F} (\eta, \vartheta) \end{array}\right\} \quad (3.60)$$

Keeping the movement parameter φ_1 constant in the equation (3.59), it is possible to express one of the other parameters to obtain a one-parameter vector–scalar function to express the contact curve for the given value φ_1.

When the explicit function among the surface parameters from the equation for contact conditions of constant φ_1 cannot be derived in explicit form then, by choosing a series of discrete values for one of the surface parameters, within the domain of variability belonging to a real tooth surface, the value of the other parameter can be calculated from equation (3.60).

Using the chosen and calculated pairs of parameters, the coordinates of contact curve points (using the equation (3.60)) can be determined point by point on the helicoid of the worm ($\vec{r}_{1F}$) and on the surface of the tool ($\vec{r}_{2F}$):

$$
\left.
\begin{aligned}
\vec{n}_{1F} \cdot \vec{v}_{1F} &= 0 \\
\vec{r}_{1F} &= \vec{r}_{1F}(\eta; \vartheta) \\
\vec{r}_2 &= \underline{\mathbf{M}}_{2,1F} \cdot \vec{r}_{1F}
\end{aligned}
\right\}
\qquad (3.61)
$$

Rotating the points of the characteristic into plane z_2, x_2 the axial section of the wheel is obtained (see Figure 3.16).

The element No2 (grinding wheel) is shaped up as the enveloping surface for the series of element No1 (the worm). The equation of element No2 in the K_{2F} system can be determined as:

$$
\vec{r}_{2F} = \underline{\mathbf{M}}_{2F,1F} \cdot \vec{r}_{1F} \qquad (3.62)
$$

The locus of contact curves in the system joined to the grinding wheel is determined by the scalar equations equivalent to vector equation (3.61):

$$
\left.
\begin{aligned}
\vec{n}_{1F} \cdot \vec{v}_{1F}^{(12)} &= 0 \\
x_{1F} &= x_{1F}(\eta, \vartheta) \\
y_{1F} &= y_{1F}(\eta, \vartheta) \\
z_{1F} &= z_{1F}(\eta, \vartheta) \\
x_{2F} &= x_{2F}(\eta, \vartheta, \varphi_1, \varphi_2) \\
y_{2F} &= y_{2F}(\eta, \vartheta, \varphi_1, \varphi_2) \\
z_{2F} &= z_{2F}(\eta, \vartheta, \varphi_1, \varphi_2)
\end{aligned}
\right\}
\qquad (3.63)
$$

A numerical example has been worked out to demonstrate the use of the method to determine the points of the characteristics (Figures 3.14 and 3.15) and the axial section of the grinding wheel (Figure 3.17). The data for the numerical example are:

$$
i=12.5, \ a=280 \text{ mm}, \ m=12 \text{ mm}, \ \rho_{ax} = 50 \text{ mm}
$$

The block diagram prepared for the computer control of the steps of the calculation can be seen in Fig. 3.17.

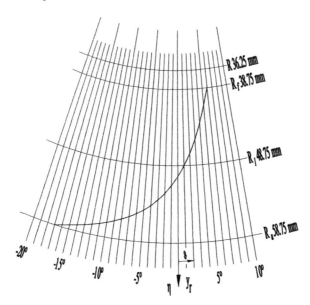

Figure 3.14 *Curve of characteristic and scale of η and ϑ coordinates valid for this example*

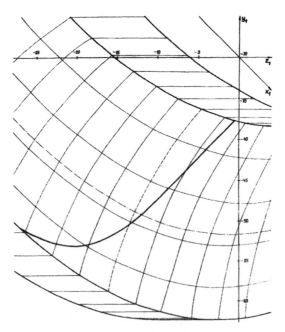

Figure 3.15 *Points of characteristic curve in K₁ coordinate system valid for this example*

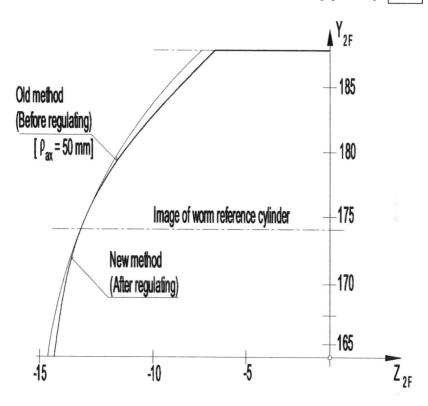

Figure 3.16 *Profile of grinding wheel in axial section using a = 223 mm centre distance for grinding*

Figure 3.17 provides the basis for preparation of the computer program which will not be presented here for reasons of length.

3.2 INVESTIGATION OF GEOMETRIC PROBLEMS IN MANUFACTURING CYLINDRICAL HELICOIDAL SURFACES HAVING CONSTANT LEAD; GENERAL MATHEMATICAL – KINEMATIC MODEL

The author's investigations were carried out using a kinematic, geometric model based on the research works of Gochman (1886), Litvin (1972), Krivenko (1967) and Perepelica (1981). In this chapter the model presented is suitable for the analysis of problems of manufacture of both cylindrical and conical worms of constant lead.

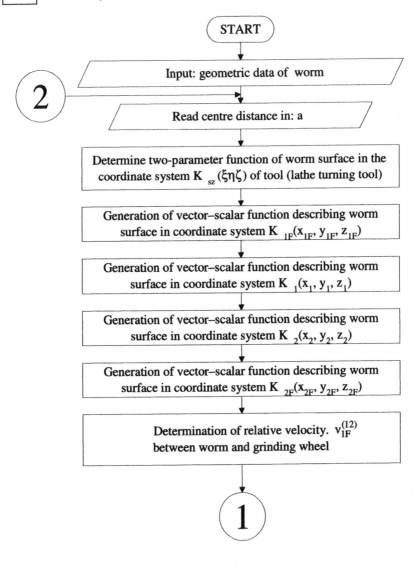

- Teeth number of worm z_1 - Centre distance a
- Module in axial section m - Reference cylinder diameter d_{01}
- Lead angle on reference   - Lead H
  cylinder γ_0 - Profile angle in axial section δ_{ax}
- Direction of tooth inclination   left   - Teeth number of worm wheel z_2
- Radius of tooth profile in axial   - Tolerance of eccentricity for
  section ρ_{ax} teething on worm f_{r1}
- Chordal thickness of tooth $\overline{S}_{n1}$ - Tolerance for axial pitch of the
- Addendum height adjust at   worm f_{p1}
  measurement of tooth chordal   - Tolerance for lead of the thread f_γ
  thickness $\overline{f}_{n1}$ - Tolerance for tooth profile of worm f_f

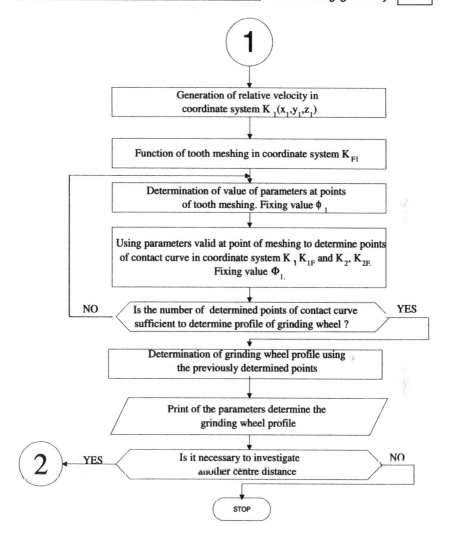

Figure 3.17 _Block diagram representing the process of determination of the equations valid for worm and worm gear having circular profile in axial section_

3.2.1 INVESTIGATION OF GEOMETRIC PROBLEMS WHEN MANUFACTURING CYLINDRICAL HELICOID SURFACES USING GENERAL MATHEMATICAL – KINEMATIC MODEL

The investigation of geometric problems related to the manufacture of cylindrical helicoid surfaces is best carried out in a general system. This general system makes it possible to discuss cylindrical worm surfaces (see Figures 3.1 and 4.4), their different manufacturing methods and generating tools.

First, $\underline{P}_{1h}$, the general transformation matrix, should be determined, thus solving the connection between the coordinate systems belonging to the generating tool and the generated surface, including the kinematics of generation. Coordinate systems used in this investigation and their relationships are shown in Figure 3.18

where:

a,b,c are the coordinates of the origin (O_2), the tool coordinate system in the K_0 coordinate system,

φ_1 is the angular displacement of the helicoid (parameter for angular displacement, for meshing and for movement),

φ_2 is the angular displacement of the tool (milling cutter or grinding wheel),

$i_{21} = \varphi_2 / \varphi_1$ gearing ratio,

γ is the inclination of the tool equal to the lead angle on the reference cylinder of the helicoid,

α is the inclination of the tool to the profile of the helicoid measured in characteristic section (eg grinding of involute worm using plane flank surface tool),

p is the lead parameter.

The coordinate systems are as follows:

- $K_0(x_0, y_0, z_0)$ stationary coordinate system joined to manufacturing machine tool;
- $K_{1F}(x_{1F}, y_{1F}, z_{1F})$ coordinate system following a thread path, joined to worm surface;
- $K_2(x_2, y_2, z_2)$ stationary coordinate system of tool;
- $K_{2F}(x_{2F}, y_{2F}, z_{2F})$ rotating coordinate system of tool;
- $K_k(x_k, y_k, z_k)$ secondary coordinate system;
- $K_s(\xi, \eta, \zeta)$ coordinate system of generator curve motion (when lathe turning it is stationary).

Conditions of movement are interpreted on the right-hand thread surfaces and on the right side of thread profiles. The statements are valid for left-hand thread surfaces.

During the author's investigations, the kinematics of generation was handled so that the helicoid surface followed a thread path and the tool surface performed a rotary motion on the left side of the thread profiles; the lead of thread and generator curve together with signs should be taken into consideration (equations (3.5), (3.8) and (3.15)).

It is necessary to determine generally valid rules for generation of the cylindrical thread surface when discussing geometric problems of manufacturing in general.

The position vector r_g of the generating curve in the coordinate system $K_s(\xi, \eta, \zeta)$ is given as a function (see Figure 3.5). This generating curve can be the edge of tool (eg in lathe turning) or the

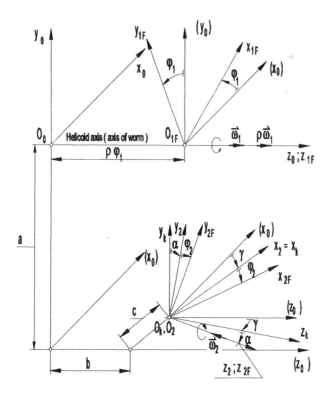

Figure 3.18 _Theory of manufacturing processes during general investigation: relative positions of cylindrical helicoid surfaces_

contact curve (eg in grinding). To formulate the equation of the generating curve, from the practical point of view let the parameter η be chosen as an independent variable.

In this way the parametric vector function of the generating curve is found to be:

$$\vec{r}_g = \xi(\eta)\vec{i} + \eta\,\vec{j} + \zeta(\eta)\vec{k} \tag{3.64}$$

The generating curve parametric equation r_g is carried by the generating curve parametric equation $K_s(\xi,\eta,\zeta)$, the coordinate system is forced on the thread path along axis z with parameter p, so the generating curve will describe a thread surface in coordinate system $K_{1F}(x_{1F}, y_{1F}, z_{1F})$ which, before this movement was performed, coincided with the K_s coordinate system (see Figure 3.19).

The thread surface described by the generating curve r_g can be determined in $K_{1F}(x_{1F}, y_{1F}, z_{1F})$ coordinate system as:

$$\vec{r}_{1F} = \underline{M}_{1F,S} \cdot r_g \tag{3.65}$$

where:

$\vec{r}_{1F}$ is the position vector of an oblique point fitted on thread surface,

$\underline{M}_{1F,S}$ is the transformation matrix between coordinate system K_s and K_{1F} (using homogeneous coordinates):

$$\underline{M}_{1F,S} = \begin{bmatrix} \cos\vartheta & -\sin\vartheta & 0 & 0 \\ \sin\vartheta & \cos\vartheta & 0 & 0 \\ 0 & 0 & 1 & p\cdot\vartheta \\ 0 & 0 & 0 & 0 \end{bmatrix} \tag{3.66}$$

Therefore, the general equation of the cylindrical worm surface is:

$$\left. \begin{aligned} x_{1F} &= \xi(\eta)\cdot\cos\vartheta - \eta\cdot\sin\vartheta \\ y_{1F} &= \xi(\eta)\cdot\sin\vartheta + \eta\cdot\cos\vartheta \\ z_{1F} &= \xi(\eta) + p\cdot\vartheta \end{aligned} \right\} \tag{3.67}$$

It can be seen from the structure of transformation matrix $\underline{M}_{1F,S}$ and the general equation of the worm surface (3.67) that the generating curve $\vec{r}_g$ and worm parameter p determine basically the worm surface. The generator curve r_g has a decisive role in the case of tool surface generation too. During generation of the tool surface,

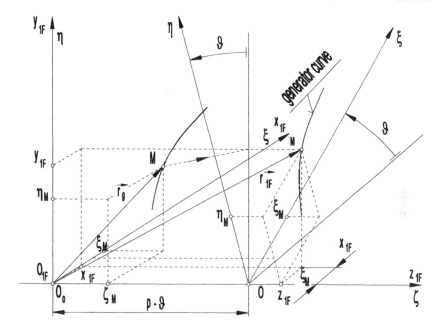

Figure 3.19 *The r_g generating curve of the thread surface in K_{1F} coordinate system*

the generator curve can be the meridian curve or the contact curve. In this case the r_{gsz} curve is interpreted in the $K_{20}(x_{20}, y_{20}, z_{20})$ coordinate system using y_{20} as a parameter, so its form is:

$$r_{gsz} = x_{20}(y_{20})\vec{i} + y_{20}j + z_{20}(y_{20})\vec{k} \qquad (3.68)$$

Rotating $\vec{r}_{gsz}$ generating curve with the $K_{20}(x_{20}, y_{20}, z_{20})$ coordinate system round the z_{20} axis, the r_{gsz} curve will describe the tool surface in the $K_{2F}(x_{2F}, y_{2F}, z_{2F})$ coordinate system (see Figure 3.20).

The tool surface determined this way could be written as:

$$\vec{r}_{2F} = \underline{M}_{2F,20} \cdot r_{gsz} \qquad (3.69)$$

where:

$\vec{r}_{2F}$ is the position vector of an oblique point fitted on tool surface,

$\underline{M}_{2F,20}$ is the transformation matrix between coordinate systems K_{2F} and K_{20}.

This matrix can be written:

$$\underline{M}_{2F,20} = \begin{bmatrix} \cos\psi & -\sin\psi & 0 & 0 \\ \sin\psi & \cos\psi & 0 & 0 \\ 0 & 0 & 1 & 0 \\ 0 & 0 & 0 & 1 \end{bmatrix} \tag{3.70}$$

Therefore, the equation of the circular symmetrical tool surface is:

$$\left. \begin{array}{l} x_{2F} = x_{20}(y_{20})\cos\psi - y_{20}\sin\psi \\ y_{2F} = x_{20}(y_{20})\sin\psi + y_{20}\cos\psi \\ z_{2F} = z_{20}(y_{20}) \end{array} \right\} \tag{3.71}$$

It should be clear from previously discussed facts that to determine any surface, be it either a worm surface or a tool surface, knowledge of its generating curve is necessary. To determine the generating curve, it is enough to know some of the meshings for the other surfaces, either the worm or the tool surface. When the generating curve is known in the coordinate system of the surface sought (eg the case of a turned worm surface) then the surface can directly be determined using either equation (3.67) or (3.71), then by transforming into the proper coordinate system, any surface can be determined.

3.2.1.1 Possible uses of the model described

This model is suitable for design of manufactured helicoidal surfaces with single or multiple edge tools of determined or undetermined edge geometry, as well as for the design of the manufacturing tools needed. After proper choice of parameters a, b, c, γ, α (shown in Figure 3.18), it is an appropriate method to determine worm surface of any special profile, even of non-standardized shape.

The worm surface is the relative kinematic wrapping surface of the manufacturing tool surface.

During their relative displacement the two surfaces remain in contact along a spatial curve. The general law for contact of elements is valid for any arbitrary point of this contact curve. It can be written as:

$$\vec{n}^{(1)} \cdot \vec{v}^{(12)} = \vec{n}^{(2)} \cdot \vec{v}^{(12)} = 0 \tag{3.72}$$

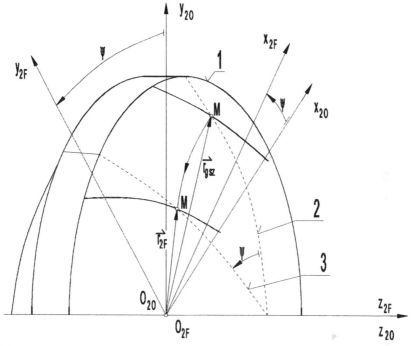

Figure 3.20 *The surface wrapped by tool generating curve (r_{gsz}) in the coordinate system K_{2F}*

1. *surface described by generation curve,*
2. *basic situation,*
3. *situation after angular displacement ψ.*

where:
$\vec{n}^{(1)}$ is normal vector of worm surface,
$\vec{n}^{(2)}$ is normal vector of tool surface,
$\vec{v}^{(1,2)}$ is vector of relative speed between surfaces worm and tool.

Knowledge of the contact curve makes possible the determination of the tool surface (*direct case*) as well as of the worm surface (*indirect case*).

Designing the tool needed to manufacture a given worm surface (direct case)

Given data is $\vec{r}_{1F} = \vec{r}_{1F}(\eta, \vartheta)$, the two-parametric vector–scalar function in the coordinate system K_{1F} (x_{1F}, y_{1F}, z_{1F}) for the surface to be generated.

Let normal vector $\vec{u}_{1F}$ be determined:

$$\vec{n}_{1F} = \frac{\partial \vec{r}_{1F}}{\partial \eta} \times \frac{\partial \vec{r}_{1F}}{\partial \vartheta} \tag{3.73}$$

The relative velocities of the two surfaces can be determined in coordinate system K_{2F} using the transformation between coordinate systems K_{1F} for worm and K_{2F} for tool:

$$\vec{v}_{2F}^{(12)} = \frac{d}{dt} \cdot \vec{r}_{2F} = \frac{d}{dt} (\underline{M}_{2F,1F}) \, \vec{r}_{\cdot 1F} \tag{3.74}$$

The vector $\vec{v}_{2F}^{(12)}$ should be transformed into coordinate system $K_{1F}(x_{1F}, y_{1F}, z_{1F})$ to determine the necessary tool surface, so:

$$\vec{v}_{1F}^{(12)} = \underline{M}_{1F,2F} \cdot \vec{v}_{2F}^{(12)} = \underline{M}_{1F,2F} \frac{d}{dt} (\underline{M}_{2F,1F}) \cdot \vec{r}_{1F} = \underline{P}_{1h} \cdot \vec{r}_{1F} \tag{3.75}$$

where the matrix for kinematic generation:

$$\underline{P}_{1h} = \underline{M}_{1F,2F} \cdot \frac{d}{dt} (\underline{M}_{2F,1F}) \tag{3.76}$$

Solving the equation (3.77) for one of its internal parameters (eg η):

$$\vec{n}_{1F}(\eta, \vartheta) \cdot \vec{v}_{1F}^{(12)}(\eta, \vartheta) = 0 \tag{3.77}$$

Applying solution:

$$\vec{r}_{1F} = \vec{r}_{1F}(\eta, \vartheta) \tag{3.78}$$

the equation of contact curve between surfaces is obtained in the form:

$$\vec{r}_{1F} = \vec{r}_{1F}(\eta(\vartheta), \vartheta) = \vec{r}_{1F}(\vartheta)$$

which is suitable for transformation of:

$$\vec{r}_{2F}(\vartheta) = \underline{M}_{2F,1F} \cdot \vec{r}_{1F}(\vartheta) \tag{3.79}$$

into the tool generating system which is the generating curve of the tool. $\underline{M}_{2F,1F}$ and $\underline{M}_{1F,2F}$ are the transformation matrices between co-ordinate systems K_{1F} and K_{2F}.

Determination of worm surface that can be manufactured using a given tool surface (indirect case)

The procedure is similar to the steps carried out in the direct case, but the direction of transformation is the opposite.

Known data:

$$\vec{r}_{2F} = \vec{r}_{2F}(y_{20}, \Psi) \tag{3.80}$$

$$\vec{n}_{2F} = \frac{\partial \vec{r}_{2F}}{\partial y_{20}} \times \frac{\partial \vec{r}_{2F}}{\partial \psi} \tag{3.81}$$

$$\vec{V}_{2F} = P_{2h} \cdot \vec{r}_{2F} \tag{3.82}$$

where the matrix for kinematic generation for inverse operation is:

$$\underline{P}_{2h} = \underline{M}_{2F,1F} \cdot \frac{d}{dt}(\underline{M}_{1F,2F}) \tag{3.83}$$

Solving the following system of equations:

$$\vec{n}_{2F} \cdot \vec{V}_{2F} = 0 \tag{3.84}$$

$$\vec{r}_{1F} = \underline{M}_{1F,2F} \cdot \vec{r}_{2F} \tag{3.85}$$

Solving these equations enables the optimum tool profile geometry to be determined.

The position vectors $\vec{r}_{2F}$ or $\vec{r}_{1F}$ describe the searched surfaces in the direct case or the indirect case; these surfaces can be generated using modern CNC machine tools or traditional machine tools supplied with additional equipment. Software for surface design (CAD) produce commands to operate CNC systems generating worm or tool surfaces using CAM systems. For example, see Figure 3.21.

While not including all the steps in the calculation, the effect of the manufacturing parameters shown in Figure 3.18 on the transformation is expressed by the matrices (3.33) and (3.43) while the kinematical conditions are shown by matrices (3.86) and (3.87).

The designed helicoidal surface, in this case the worm in its axial section with circular arched profile, or the wrapping worm of the generating milling cutter is shown in Figure 3.21a.

Machining using the disc-like tool shown in Figure 3.21a, the characteristic is shaped up in polar coordinates as can be seen in Figure 3.21b. Using this characteristic the axial section of tool (see Figure 3.21c) can be determined.

The results of the calculations can be directly utilized when designing a grinding wheel or circular milling cutter or with given manufacturing kinematics to determine an arbitrary tool profile for use as input data for manufacturing on a CNC machine tool.

An example of such applications can be seen in Figures 4.7, 4.8, 4.12 and in Figures 4.36 and 4.37.

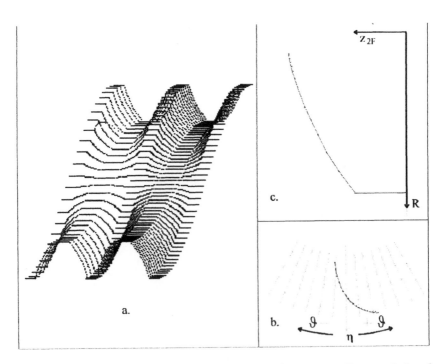

Figure 3.21 *The designed worm surface, a, the contact characteristics, b, and the axial section of the generating tool (grinding wheel or circular milling cutter), c*

$$\underline{P}_{1h} = M_{1F,2F} \cdot \frac{d}{dt} M_{2F,1F} =$$

$$
\begin{bmatrix}
0 & -(1+i\cdot\cos\alpha\cdot\cos\gamma) & +i\cdot\sin\varphi_1\cdot\cos\alpha\cdot\sin\gamma - i\cdot\cos\varphi_1\cdot\sin\alpha & \cos\varphi_1\cdot(-i\cdot p\cdot\varphi_1\cdot\sin\alpha - i\cdot a\cdot\cos\alpha\cdot\cos\gamma + i\cdot b\cdot\sin\alpha\cdot\cos\gamma + i\cdot c\cdot\sin\gamma\cdot\sin\alpha) + \sin\varphi_1\cdot(i\cdot p\cdot\varphi_1\cdot\cos\alpha\cdot\sin\gamma - i\cdot c\cdot\cos\alpha) \\[2ex]
1+i\cdot\cos\alpha\cdot\cos\gamma & 0 & +i\cdot\cos\varphi_1\cdot\cos\alpha\sin\gamma + i\cdot\sin\varphi_1\cdot\sin\alpha & \sin\varphi_1\cdot(i\cdot p\cdot\varphi_1\cdot\sin\alpha + i\cdot a\cdot\cos\gamma\cdot\cos\alpha - i\cdot b\cdot\cos\gamma\sin\alpha - i\cdot c\cdot\sin\gamma\cdot\sin\alpha) + \sin\varphi_1\cdot(i\cdot p\cdot\varphi_1\cdot\cos\alpha\cdot\sin\gamma - i\cdot c\cdot\cos\alpha) \\[2ex]
-i\cdot\sin\gamma\cdot\sin\varphi_1\cdot\cos\alpha + i\cdot\cos\varphi_1\cdot\sin\alpha & -i\cdot\sin\alpha\cdot\sin\varphi_1 - i\cdot\cos\alpha\cdot\sin\gamma\cdot\cos\varphi_1 & 0 & p - i\cdot a\cdot\sin\gamma\cdot\cos\alpha - i\cdot c\cdot\cos\gamma\cdot\sin\alpha + i\cdot b\cdot\sin\alpha\cdot\sin\gamma \\[2ex]
0 & 0 & 0 & 0
\end{bmatrix}
\tag{3.86}
$$

$$P_{2h} = \underline{M}_{2F,1F} \cdot \frac{d}{dt}\underline{M}_{1F,2F} =$$

$$
\begin{bmatrix}
0 & i+\cos\alpha\cdot\cos\gamma & \begin{array}{c}\sin\varphi_2\cdot\sin\gamma- \\ -\cos\varphi_2\cdot\sin\alpha\cdot\cos\gamma\end{array} & \begin{array}{c}\sin\varphi_2\cdot\sin\alpha\cdot(-a\cdot\sin\gamma+p\cdot\cos\gamma)+ \\ +\cos\varphi_2\cdot(-p\cdot\sin\gamma-a\cdot\cos\gamma)+ \\ +c\cdot\sin\varphi_2\cdot\cos\alpha\end{array} \\[4ex]
-1-\cos\alpha\cdot\cos\gamma & 0 & \begin{array}{c}\cos\varphi_2\cdot\sin\gamma- \\ -\sin\varphi_2\cdot\cos\gamma\cdot\sin\alpha\end{array} & \begin{array}{c}\cos\varphi_2\cdot\sin\alpha\cdot(a\cdot\sin\gamma-p\cdot\cos\gamma)+ \\ +\sin\varphi_2\cdot(-p\cdot\sin\gamma-a\cdot\cos\gamma)- \\ +c\cdot\cos\varphi_2\cdot\cos\alpha\end{array} \\[4ex]
\begin{array}{c}\sin\gamma\cdot\sin\varphi_2+ \\ +\cos\varphi_2\cdot\sin\alpha\cdot\cos\gamma\end{array} & \begin{array}{c}-\sin\gamma\cdot\cos\varphi_2+ \\ +\cos\gamma\cdot\sin\alpha\cdot\sin\varphi_2\end{array} & 0 & \begin{array}{c}\cos\alpha\cdot(a\cdot\sin\gamma-p\cdot\cos\gamma)+ \\ +c\cdot\sin\alpha\end{array} \\[4ex]
0 & 0 & 0 & 0
\end{bmatrix}
$$

$$(3.87)$$

3.2.2 Analysis of manufacturing geometry for conical helicoid surfaces

Conical helicoid surfaces can be frequently used in practice, but the most frequent use is as the active surface of the conical worm.

The spiroid drive, consisting of a conical worm mated with a face gear, can be used with advantage as a drive for robotic mechanisms, for machine tool drives having no clearance etc. Clearance-free operation can be achieved by simple axial displacement of the worm (Bohle, 1956).

It should be mentioned that when manufacturing spiroid drives, the grinding, the last step, is not widely carried out because to determine the grinding wheel profile needed to produce conical worm surfaces with the required precision is not an easy task (Boecker and Rochel, 1965; Bohle and Saari, 1995). That is why the majority of textbooks mention this drive but no detailed analysis concerning manufacturing geometry can be found.

The aim of this chapter is to make clear the correct manufacturing geometry of conical worm surfaces and to provide a method for the design of grinding, being the finishing process of machining, with the precision needed to fulfil practical requirements.

Two problems arise when finishing conical worm surfaces with a grinding wheel, which make it practically impossible to produce a profile having precise geometry:

1. During grinding, owing to the wear of the grinding wheel, both diameter and profile of the tool change. This is why the generated surface is a deformed one compared with the theoretical profile and to the profile generated in the machining process.
2. The conical worm surface diameter is changing axially, which itself creates a continuous change in worm profile (for constant profile of the grinding wheel).

Case 1. occurs during the grinding of cylindrical worm surfaces also, as in the case of a constant wheel profile, the profile is a function of the distance between the centre lines of the helicoidal surface and the grinding wheel and the ratio of diameters of the worm and the grinding wheel. This problem can only be solved by using a superhard grinding wheel having nearly no wear or by applying a regulating process and checking the helicoid surface often to control it, usually when grinding using an alundum wheel. In this second case, owing to time-consuming regulations, the machining process is less efficient and because of the ever-changing position of the workpiece the possibility of mistakes accumulates.

In case 2. this problem, owing to the nature of conical worm surface geometry, can only be solved by continuous regulation of the wheel profile to produce a correct worm surface at any arbitrary diameter along its axis.

To realize this theoretical possibility, NC- or CNC-controlled wheel-regulating devices should be connected to the grinding wheel. From an economic point of view, this solution is not suitable because of the high measure of wear of the grinding wheel. Using the circular milling cutter is not a solution.

The problems mentioned before, in manufacturing a hob for machining face gear, the other element of spiroid drive, remain.

Instead of the theoretically exact continuous wheel regulation it would be useful to choose another regulation method fulfilling the practical requirement. This way can only be accepted if, when using constant wheel profile, it is possible to keep the values of the characteristics of worm profile in both the sections, at minor and at major diameters of the worm, within allowable limits.

When manufacturing either a spiroid worm or a spiroid generator milling cutter, a basic problem is to determine the position of generation of the grinding wheel. The author's suggestion to solve this problem is that the advantageous position along the axis of conical worm surface to generate the grinding wheel profile is chosen according to the optimal value of kinematical parameter φ_1.

This chapter aims to present a mathematical model for general use in manufacturing spiroid drives for design of tool and machining geometry. To justify the efficacy of this method in practice, a number of drives were manufactured and tested (see Figures 4.43, 5.13 and 5.14).

3.2.2.1 Types of conical helicoid surfaces

The tooth surface of a conical worm, as an element of a spiroid drive, can be similarly generated as with cylindrical worms, but linked to axial displacement of the tool (p_a), depending on the measure of conical shape of the worm, a tangential displacement (p_t) of the tool is necessary as well. All this can be seen in Figures 3.22, 3.23 and 3.24 which show the types of helicoids and their equations too. Similarly to cylindrical worms with ruled surfaces, on the surfaces of spiroid worms several types of helicoids – involute, Archimedian or convolute – can be differentiated. But non-ruled conical worm surfaces can be envisaged too (see Figure 1.1 and Leroy _et al_, 1986).

The teething of face gear is generated, according to present practice, using a worm hob having a surface equivalent to a worm wrapping surface of a conical worm. This method is called _direct generation_ in the literature. The involute worm gear drive with ground ruled surface (Bluzat, 1986; Gransin, 1972) has a great advantage, namely the identity of surfaces can be realized by simple machining technology. When the thread surface of the worm or worm gear milling cutter is machined using a conical or ring shape grinding wheel, the precision of the profile surfaces cannot be easily guaranteed (Bluzat, 1986).

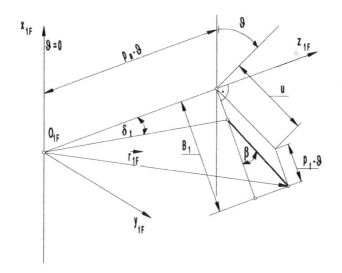

Figure 3.22 _Generator of conical Archimedian helicoid in oblique position_

The equation of conical Archimedian helicoid surface is:

$$\vec{r}_{1F} = \begin{bmatrix} -B_1 \cdot \sin \vartheta \\ +B_1 \cdot \cos \vartheta \\ u \cdot \sin \beta + p_a \cdot \vartheta \\ 1 \end{bmatrix} \tag{3.88}$$

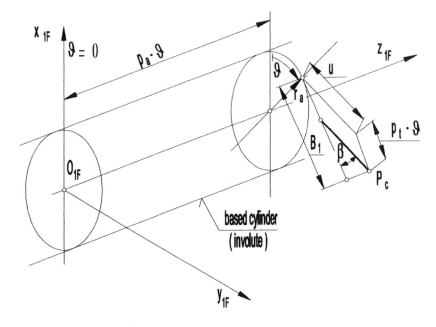

Figure 3.23 *Generator of conical involute helicoidal surface in oblique position*

The equation of conical involute helicoidal surface is:

$$\vec{r}_{1F} = \begin{bmatrix} -B_1 \cdot \sin\vartheta + r_a \cdot \cos\vartheta \\ B_1 \cdot \cos\vartheta + r_a \cdot \sin\vartheta \\ u \cdot \sin\beta + p_a \cdot \vartheta \\ 1 \end{bmatrix} \tag{3.89}$$

where $r_a = p_a \cdot \operatorname{ctg}\beta - p_t$

Regarding Figures 3.22, 3.23 and 3.24, it is possible to write:

$$B_1 = u.\cos\beta + p_t .\vartheta \tag{3.90}$$

$$p_t = p_a.\operatorname{tg}\delta_1 \tag{3.91}$$

where δ_1 is the characteristic of conical shape.

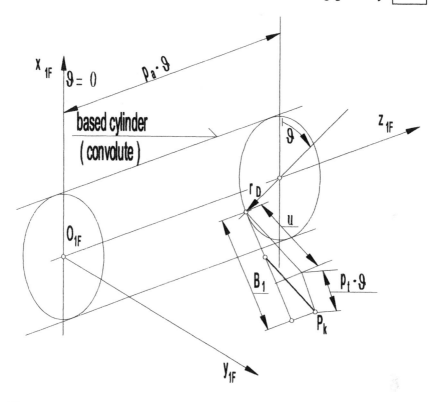

Figure 3.24 _Generator of conical convolute helicoidal surface in oblique position_

Equation of conical convolute helicoidal surface is:

$$\vec{r}_{1F} = \begin{bmatrix} -B \cdot \sin \vartheta - r_t \cdot \cos \vartheta \\ B_1 \cdot \cos \vartheta - r_t \cdot \sin \vartheta \\ u \cdot \sin \beta + p_a \cdot \vartheta \\ 1 \end{bmatrix} \tag{3.92}$$

General ruled surface on conical worm

Using a suitable equation to define any oblique point fitted on any type of conical worm surface, such a formula for the position vector is obtained as can be suitable to describe the generally valid form for all the three types of conical worm surfaces (Figure 3.25):

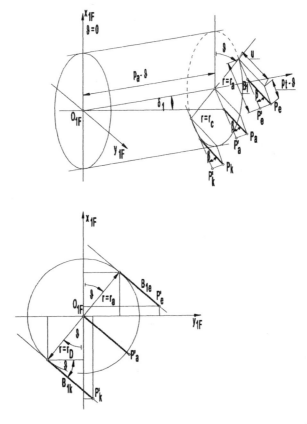

Figure 3.25 *Summary of generation of ruled conical worm surfaces*

$$\vec{r}_{1F} = \begin{bmatrix} -B_1 \cdot \sin \vartheta + r \cdot \cos \vartheta \\ B_1 \cdot \cos \vartheta + r \cdot \sin \vartheta \\ u \cdot \sin \beta + p_a \cdot \vartheta \\ 1 \end{bmatrix} \tag{3.93}$$

The above generally valid formula defines:

$$\left. \begin{array}{l} \text{in case of } r = 0 \text{ Archimedian} \\ r = r_a = p_a \times \mathrm{ctg}\, \beta - p_t > 0 \text{ involute} \\ -r_a < r = r_0 < \text{convolute helicoid surface} \end{array} \right\} \tag{3.94}$$

For a helicoidal surface, by substituting $\delta_1 = 0$ in all three cases, the corresponding cylindrical worm is obtained.

3.2.2.2 Kinematic analysis of machining, using axisymmetrical tool, of contact surfaces situated over conical surface

The kinematic conditions for machining can be described by the relative position of coordinate systems as shown in Figure 3.26.

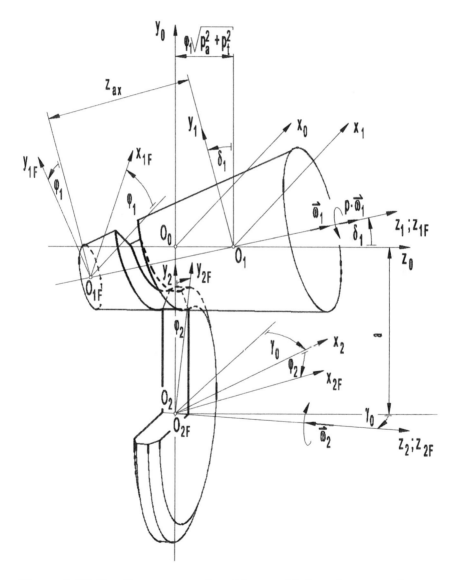

Figure 3.26 _Coordinate systems describing machining of worm surfaces having conical addendum and dedendum acting surfaces_

To analyse kinematic conditions, first interpret the movements of coordinate systems themselves. The coordinate system K_{1F} rotates with angular speed:

$$\omega_1 = \frac{d\varphi_1}{dt} = \text{constant} \qquad (3.95)$$

in coordinate system K_1.

The coordinate system K_1 performs translational movement in coordinate system K_o. The distance between origins O_1 and O_o is:

$$\overline{O_1 O_o} = \varphi_1 \cdot \sqrt{p_a^2 + p_t^2} \qquad (3.96)$$

It is linearly proportional to the angular displacement φ_1 of worm surfaces in coordinate system K_1, and

$$\sqrt{p_a^2 + p_t^2} = \text{constant} \qquad (3.97)$$

where:

p_a = axial displacement of tool,
p_t = constant, equal to the tangential displacement that has already been applied in equation (3.93) describing the worm surface. It is characteristic of conical shape (see Figure 3.25).

The coordinate system K_2 is joined to a suitable point of the pitch-cone surface round axis y_2, the point is optimal to φ_1, and is inclined with angle γ_o to the lead angle belonging to that assigned point. $\varphi_{1\text{opt.}}$ is the angle that guarantees the minimum inclination of shape, the value of the meshing parameter.

The coordinate system K_{2F} rotates in coordinate system K_2 with constant angular speed:

$$\omega_2 = \frac{d\varphi_2}{dt} = \text{constant} \qquad (3.98)$$

The transformation matrices between the coordinate systems are as follows:

$$(3.99)$$

$$M_{1,1F} = \begin{bmatrix} \cos\varphi_1 & -\sin\varphi_1 & 0 & 0 \\ \sin\varphi_1 & \cos\varphi_1 & 0 & 0 \\ 0 & 0 & 1 & -z_{ax} \\ 0 & 0 & 0 & 1 \end{bmatrix} \quad M_{1F,1} = \begin{bmatrix} \cos\varphi_1 & \sin\varphi_1 & 0 & 0 \\ -\sin\varphi_1 & \cos\varphi_1 & 0 & 0 \\ 0 & 0 & 1 & +z_{ax} \\ 0 & 0 & 0 & 1 \end{bmatrix}$$

$$\underline{M}_{0,1} = \begin{bmatrix} 1 & 0 & 0 & 0 \\ 0 & \cos ä_1 & \sin ä_1 & 0 \\ 0 & -\sin ä_1 & \cos ä_1 & \varphi_1 \cdot \sqrt{p_a^2 + p_t^2} \\ 0 & 0 & 0 & 1 \end{bmatrix}$$

$$\underline{M}_{1,0} = \begin{bmatrix} 1 & 0 & 0 & 0 \\ 0 & \cos ä_1 & -\sin ä_1 & \varphi_1 \cdot \sin ä_1 \cdot \sqrt{p_a^2 + p_t^2} \\ 0 & \sin ä_1 & \cos ä_1 & -\varphi_1 \cdot \cos ä_1 \cdot \sqrt{p_a^2 + p_t^2} \\ 0 & 0 & 0 & 1 \end{bmatrix} \qquad (3.100)$$

$$\underline{M}_{2,0} = \begin{bmatrix} \cos ã & 0 & \sin ã & 0 \\ 0 & 1 & 0 & a \\ -\sin ã & 0 & \cos ã & 0 \\ 0 & 0 & 0 & 1 \end{bmatrix} \qquad \underline{M}_{0,2} = \begin{bmatrix} \cos ã & 0 & -\sin ã & 0 \\ 0 & 1 & 0 & -a \\ \sin ã & 0 & \cos ã & 0 \\ 0 & 0 & 0 & 1 \end{bmatrix} \qquad (3.101)$$

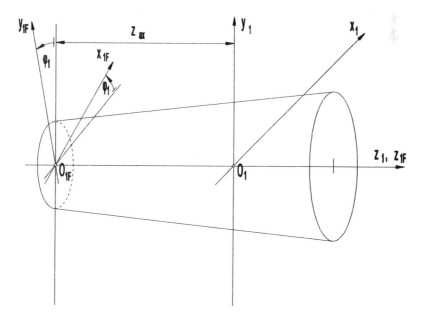

Figure 3.27 *Connection between coordinates (x_1, y_1, z_1) joined to the machine tool and coordinates (x_{1F}, y_{1F}, z_{1F}) joined to the rotating worm systems*

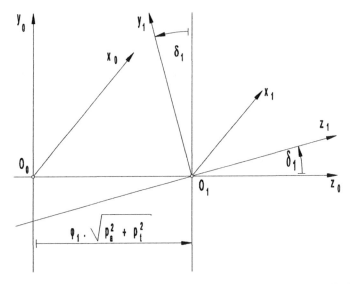

Figure 3.28 *Connection between stationary coordinate systems* K_o (x_o, y_o, z_o) *and* K_1 (x_1, y_1, z_1) *joined to table of machine tool*

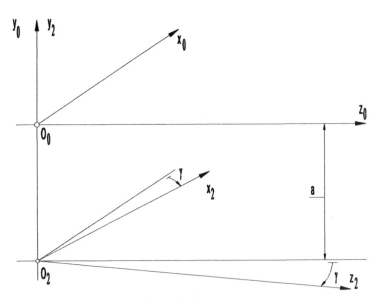

Figure 3.29 *Connection of coordinate systems* K_o (x_o, y_o, z_o) *and* K_2 (x_2, y_2, z_2)

$$
M_{2F,2} = \begin{bmatrix} \cos\varphi_2 & -\sin\varphi_2 & 0 & 0 \\ \sin\varphi_2 & \cos\varphi_2 & 0 & 0 \\ 0 & 0 & 1 & 0 \\ 0 & 0 & 0 & 1 \end{bmatrix} \quad M_{2,2F} = \begin{bmatrix} \cos\varphi_2 & \sin\varphi_2 & 0 & 0 \\ -\sin\varphi_2 & \cos\varphi_2 & 0 & 0 \\ 0 & 0 & 1 & 0 \\ 0 & 0 & 0 & 1 \end{bmatrix} \quad (3.102)
$$

Knowing the transformation matrices, to determine the position vector of any oblique point in any coordinate system, the position vector in two-parameter form is given as:

$$
\vec{r}_{1F} = \vec{r}_{1F}(u,\vartheta) = \begin{bmatrix} x_{1F}(u,\vartheta) \\ y_{1F}(u,\vartheta) \\ z_{1F}(u,\vartheta) \\ 1 \end{bmatrix} \quad (3.103)
$$

The position vectors of the same point in other systems are:

$$
\left. \begin{aligned}
\vec{r}_1 &= \underline{M}_{1,1F} \cdot \vec{r}_{1F} \\
\vec{r}_0 &= \underline{M}_{0,1F} \cdot \vec{r}_{1F} \quad \text{where} \quad \underline{M}_{0,1F} = \underline{M}_{0,1} \cdot \underline{M}_{1,1F} \\
\vec{r}_2 &= \underline{M}_{2,1F} \cdot \vec{r}_{1F} \qquad \text{where} \quad \underline{M}_{2,1F} = \underline{M}_{2,0} \cdot \underline{M}_{0,1} \cdot \underline{M}_{1,1F} \\
\vec{r}_{2F} &= \underline{M}_{2F,1F} \cdot \vec{r}_{1F} \quad \text{where} \quad \underline{M}_{2F,1F} = \underline{M}_{2F,2} \cdot \underline{M}_{2,0} \cdot \underline{M}_{0,1} \cdot \underline{M}_{1,1F}
\end{aligned} \right\} \quad (3.104)
$$

Later on, transformation matrices between rotating systems will be determined as $\underline{M}_{1F,2F}$ and $\underline{M}_{2F,1F}$. The matrix $\underline{M}_{1F,2F}$ is given by equation (3.111) while $\underline{M}_{2F,1F}$ by equation (3.112).

When seeking a surface in coordinate system K_{2F} suitable to mate with the surface characterized by position vector $\vec{r}_{1F} = \vec{r}_{1F}(u,\vartheta)$ in coordinate system K_{1F} the fact that surfaces mated and enveloping each other during their common movement fulfil the requirement:

$$
\varphi_2 = i_{21} \times \varphi_1 \quad (3.105)
$$

and the nature of the enveloping can be expressed by movement parameter (φ_1). Being the surface No2, ie the meshing surface of machining tool, it is in practice a block of rotation. So further investigation should be limited to determination of the momentary contact curve or in other words the contact characteristics of elements No1 and No2, the practical geometric shape of element No2 being unambiguously determined by the characteristics.

According to the law of contact:

$$\vec{n}_{1F}\cdot\vec{V}_{1F}^{(12)}=\vec{n}_{2F}\cdot\vec{V}_{2F}^{(12)}=\vec{n}\cdot\vec{V}^{(12)}=0 \tag{3.106}$$

The internal parameters (u, ϑ) satisfying the equation should be substituted into the vector–scalar function equation of the surface No1, valid in the coordinate system K_{1F}.

Differentiating with respect to time, the relative velocity and relative position of the two blocks:

$$\left.\begin{array}{l} \vec{r}_{2F} = \underline{\mathbf{M}}_{2F,1F}\cdot\vec{r}_{1F} \\[2mm] \dot{\varphi}_2 = i_{21}\cdot\dot{\varphi}_1 \end{array}\right\} \tag{3.107}$$

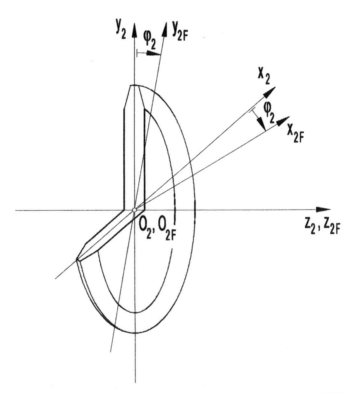

Figure 3.30 *Connection of coordinate systems K_2 (x_2, y_2, z_2) and K_{2F} (x_{2F}, y_{2F}, z_{2F})*

we obtain:

$$\vec{V}_{2F}^{(12)} = \vec{r}_{2F} = \frac{d}{dt}(\underline{M}_{2F,1F}\cdot\vec{r}_{1F}) = \frac{d\underline{M}_{2F,1F}}{dt}\cdot\vec{r}_{1F} \qquad (3.108)$$

as $\vec{r}_{1F}$ is constant in time. The surface No1 is given primarily in the coordinate system K_{1F}, so to determine equation (3.39) the relative velocity vector should first be transformed from system K_{2F} into system K_{1F} using the equation:

$$\vec{V}_{1F}^{(12)} = \underline{M}_{1F,2F}\cdot\vec{V}_{2F}^{(12)} \qquad (3.109)$$

Finally, the relative velocity vector can be expressed in system K_{1F} as:

$$\vec{V}_{1F}^{(12)} = \underline{M}_{1F,2F}\cdot\frac{d\underline{M}_{2F,1F}}{dt}\cdot\vec{r}_{1F} = \underline{P}_{1k}\cdot\vec{r}_{1F} \qquad (3.110)$$

Find for the two given coordinate systems the matrices:

$$\frac{d\underline{M}_{2F,1F}}{dt} \qquad \text{and} \qquad \underline{P}_{1k} = \underline{M}_{1F,2F}\cdot\frac{d\underline{M}_{2F,1F}}{dt}$$

To simplify the mathematical formulae we shall introduce symbol i for i_{21} as in equations (3.113) and (3.114). Note finally that the matrices of kinematical generation are $\underline{P}_{h1}$ for cylindrical and p_{1k} for conical worm surfaces (3.113).

3.2.2.3 Generation of ruled conical helicoid surface: manufacture of cutting tool profile

Investigation of the kinematic problems relating to machining was carried out within given geometric conditions. The driving element was the worm (No1), which was determined by the two-parameter vector–scalar function:

$$\vec{r}_{1F} = \vec{r}_{1F}(u, \vartheta) \qquad (3.115)$$

In the kinematic model, for the manufacturing geometry of the $\vec{r}_{1F}$ surface, the basic characteristics of thread, the lead and cone angle, are relatively strictly determined, but the characteristics of the thread axial section can vary within a wide range in practice. Later on, the investigation for a general conical helicoid determined with a well-defined surface will be described (see equation (3.93)).

$$
\underline{M}_{1F,2F} = \underline{M}_{1F,0} \cdot \underline{M}_{0,2F} =
$$

$$
\begin{bmatrix}
\begin{array}{l}\cos\gamma_0\cdot\cos\varphi_1\cdot\cos\varphi_2 - \\ -\cos\delta_1\cdot\sin\varphi_1\cdot\sin\varphi_2 - \\ -\sin\gamma_0\cdot\sin\delta_1\cdot\sin\varphi_1\cdot \\ \cdot\cos\varphi_2\end{array}
&
\begin{array}{l}\cos\gamma_0\cdot\cos\varphi_1\cdot\sin\varphi_2 + \\ +\cos\delta_1\cdot\sin\varphi_1\cdot\cos\varphi_2 - \\ -\sin\gamma_0\cdot\sin\delta_1\cdot\sin\varphi_1\cdot \\ \cdot\sin\varphi_2\end{array}
&
\begin{array}{l}-\sin\gamma_0\cdot\cos\varphi_1 - \\ -\cos\gamma_0\cdot\sin\delta_1\cdot\sin\varphi_1\end{array}
&
\begin{array}{l}-a\cdot\cos\delta_1\cdot\sin\varphi_1 + \\ +\sin\delta_1\cdot\sqrt{p_a^2+p_t^2}\cdot\varphi_1\cdot\sin\varphi_1\end{array}
\\[4ex]
\begin{array}{l}-\cos\gamma_0\cdot\sin\varphi_1\cdot\cos\varphi_2 - \\ -\cos\delta_1\cdot\cos\varphi_1\cdot\sin\varphi_2 - \\ -\sin\gamma_0\cdot\sin\delta_1\cdot\cos\varphi_1\cdot \\ \cdot\cos\varphi_2\end{array}
&
\begin{array}{l}-\cos\gamma_0\cdot\sin\varphi_1\cdot\sin\varphi_2 + \\ +\cos\delta_1\cdot\cos\varphi_1\cdot\cos\varphi_2 - \\ -\sin\gamma_0\cdot\sin\delta_1\cdot\cos\varphi_1\cdot \\ \cdot\sin\varphi_2\end{array}
&
\begin{array}{l}\sin\gamma_0\cdot\sin\varphi_1 - \\ -\cos\gamma_0\cdot\sin\delta_1\cdot\cos\varphi_1\end{array}
&
\begin{array}{l}-a\cdot\cos\delta_1\cdot\cos\varphi_1 + \\ +\sin\delta_1\cdot\sqrt{p_a^2+p_t^2}\cdot\varphi_1\cdot\cos\varphi_1\end{array}
\\[4ex]
\begin{array}{l}-\sin\delta_1\cdot\sin\varphi_2 + \\ +\sin\gamma_0\cdot\cos\delta_1\cdot\cos\varphi_2\end{array}
&
\begin{array}{l}\sin\delta_1\cdot\cos\varphi_2 + \\ +\sin\gamma_0\cdot\cos\delta_1\cdot\cos\varphi_2\end{array}
&
\cos\gamma_0\cdot\cos\delta_1
&
\begin{array}{l}-a\cdot\sin\delta_1 - \\ -\cos\delta_1\cdot\sqrt{p_a^2+p_t^2}\cdot\varphi_1 - z_{ax}\end{array}
\\[4ex]
0 & 0 & 0 & 0
\end{bmatrix}
\tag{3.111}
$$

$$\underline{M}_{2F,1F} = \underline{M}_{2F,0} \cdot \underline{M}_{0,1F} =$$

$$
\begin{bmatrix}
\begin{aligned}&\cos\gamma_0 \cdot \cos\varphi_1 \cdot \cos\varphi_2 - \\ &\cos\delta_1 \cdot \sin\varphi_1 \cdot \sin\varphi_2 - \\ &\sin\gamma_0 \cdot \sin\delta_1 \cdot \cos\varphi_2 \cdot \\ &\sin\varphi_1\end{aligned}
& \begin{aligned}&-\cos\gamma_0 \cdot \sin\varphi_1 \cdot \cos\varphi_2 - \\ &\cos\delta_1 \cdot \cos\varphi_1 \cdot \sin\varphi_2 - \\ &\sin\gamma_0 \cdot \sin\delta_1 \cdot \cos\varphi_1 \cdot \\ &\cos\varphi_2\end{aligned}
& \begin{aligned}&-\sin\delta_1 \cdot \sin\varphi_2 + \\ &\sin\gamma_0 \cdot \cos\delta_1 \cdot \cos\varphi_2\end{aligned}
& \begin{aligned}&+z_{ax} \cdot \sin\delta_1 \cdot \sin\varphi_2 - \\ &z_{ax} \cdot \cos\delta_1 \cdot \sin\gamma_0 \cdot \sin\varphi_1 + \\ &\sin\gamma_0 \cdot \sqrt{p_a^2 + p_t^2} \cdot \varphi_1 \cdot \cos\varphi_2 - \\ &a \cdot \sin\varphi_2\end{aligned} \\[2em]
\begin{aligned}&\cos\gamma_0 \cdot \cos\varphi_1 \cdot \sin\varphi_2 + \\ &\cos\delta_1 \cdot \sin\varphi_1 \cdot \cos\varphi_2 - \\ &\sin\gamma_0 \cdot \sin\delta_1 \cdot \sin\varphi_1 \cdot \\ &\sin\varphi_2\end{aligned}
& \begin{aligned}&-\cos\gamma_0 \cdot \sin\varphi_1 \cdot \sin\varphi_2 + \\ &\cos\delta_1 \cdot \cos\varphi_1 \cdot \cos\varphi_2 - \\ &\sin\gamma_0 \cdot \sin\delta_1 \cdot \cos\varphi_1 \cdot \\ &\sin\varphi_2\end{aligned}
& \begin{aligned}&\sin\delta_1 \cdot \cos\varphi_2 + \\ &\sin\gamma_0 \cdot \cos\delta_1 \cdot \sin\varphi_2\end{aligned}
& \begin{aligned}&-z_{ax} \cdot \sin\delta_1 \cdot \cos\varphi_2 - \\ &z_{ax} \cdot \sin\gamma_0 \cdot \cos\delta_1 \cdot \sin\varphi_2 + \\ &\sin\gamma_0 \cdot \sqrt{p_a^2 + p_t^2} \cdot \varphi_1 \cdot \sin\varphi_2 + \\ &a \cdot \cos\varphi_2\end{aligned} \\[2em]
\begin{aligned}&-\sin\gamma_0 \cdot \cos\varphi_1 - \\ &\cos\gamma_0 \cdot \sin\delta_1 \cdot \sin\varphi_1\end{aligned}
& \begin{aligned}&-\sin\gamma_0 \cdot \sin\varphi_1 - \\ &\cos\gamma_0 \cdot \sin\delta_1 \cdot \cos\varphi_1\end{aligned}
& \cos\gamma_0 \cdot \cos\delta_1
& \begin{aligned}&-z_{ax} \cdot \cos\gamma_0 \cdot \cos\delta_1 + \\ &\cos\gamma_0 \cdot \sqrt{p_a^2 + p_t^2} \cdot \varphi_1\end{aligned} \\[2em]
0 & 0 & 0 & 0
\end{bmatrix}
$$

$$(3.112)$$

$$\frac{d}{dt}\underline{M}_{2,F_1,F} =$$

$$
\begin{bmatrix}
\begin{aligned}
&(-\cos\gamma_0 - i\cdot\cos\delta_1)\sin\varphi_1\cdot\cos\varphi_2 - \\
&(i\cdot\cos\gamma_0 + \cos\delta_1)\cos\varphi_1\cdot\sin\varphi_2 - \\
&\sin\gamma_0\sin\delta_1(\cos\varphi_1\cdot\cos\varphi_2 - \\
&i\cdot\sin\varphi_1\cdot\sin\varphi_2)
\end{aligned}
&
\begin{aligned}
&(\cos\gamma_0 - i\cdot\cos\delta_1)\cos\varphi_1\cdot\cos\varphi_2 + \\
&(i\cdot\cos\gamma_0 + \cos\delta_1)\sin\varphi_1\cdot\sin\varphi_2 + \\
&\sin\gamma_0\sin\delta_1(\sin\varphi_1\cdot\cos\varphi_2 + \\
&i\cdot\cos\varphi_1\cdot\sin\varphi_2)
\end{aligned}
&
\begin{aligned}
&-i\cdot\sin\delta_1\cdot\cos\varphi_2 - \\
&i\cdot\sin\gamma_0\cos\delta_1\cdot\sin\varphi_2
\end{aligned}
&
\begin{aligned}
&z_{ax}i\cdot\cos\delta_1\cdot\sin\gamma_0\cdot\sin\varphi_2 - \\
&(z_{ax}i\cdot\sin\delta_1 + a\cdot i - \\
&\sqrt{p_a^2+p_i^2}\cdot\sin\gamma_0)\cos\varphi_2 - \\
&i\cdot\sin\gamma_0\cdot\sqrt{p_a^2+p_i^2}\cdot\\
&\varphi_1\cdot\sin\varphi_2
\end{aligned}
\\[2em]
\begin{aligned}
&(-\cos\gamma_0 - i\cdot\cos\delta_1)\sin\varphi_1\cdot\sin\varphi_2 + \\
&(i\cdot\cos\gamma_0 + \cos\delta_1)\cos\varphi_1\cdot\cos\varphi_2 + \\
&i\cdot\sin\varphi_1 - \\
&\sin\gamma_0\sin\delta_1(\cos\varphi_1\cdot\sin\varphi_2 + i\cdot\sin\varphi_1\\
&\cdot\cos\varphi_2)
\end{aligned}
&
\begin{aligned}
&(-\cos\gamma_0 - i\cdot\cos\delta_1)\cos\varphi_1\cdot\sin\varphi_2 - \\
&(i\cdot\cos\gamma_0 + \cos\delta_1)\sin\varphi_1\cdot\sin\varphi_2 + \\
&\sin\gamma_0\sin\delta_1(\sin\varphi_1\cdot\sin\varphi_2 - \\
&i\cdot\cos\varphi_1\cdot\cos\varphi_2)
\end{aligned}
&
\begin{aligned}
&-i\cdot\sin\delta_1\cdot\sin\varphi_2 + \\
&i\cdot\sin\gamma_0\cos\delta_1\cdot\sin\varphi_2;
\end{aligned}
&
\begin{aligned}
&-z_{ax}i\cdot\sin\delta_1 + \sqrt{p_a^2+p_i^2}\cdot \\
&\sin\gamma_0 - (\sin\gamma_0 - a\cdot i)\cdot\sin\varphi_2 + \\
&z_{ax}\cdot i\cdot\sin\gamma_0\cdot\cos\delta_1\cdot \\
&\cos\delta_1\cdot\cos\varphi_2 + \\
&i\sqrt{p_a^2+p_i^2}\cdot\sin\gamma_0\cdot \\
&\varphi_1\cdot\cos\varphi_2
\end{aligned}
\\[2em]
\sin\gamma_0\cdot\sin\varphi_1 - \cos\gamma_0\cdot\sin\delta_1\cdot\cos\varphi_1
&
\sin\gamma_0\cdot\cos\varphi_1 + \cos\gamma_0\cdot\sin\delta_1\cdot\sin\varphi_1
&
0
&
\sqrt{p_a^2+p_i^2}\cdot\cos\gamma_0
\end{bmatrix}
$$

$$(3.113)$$

$$\underline{P}_{1k} = \underline{M}_{1F,2F} \cdot \frac{d}{dt} \cdot \underline{M}_{2F,1F} =$$

$$
\begin{bmatrix}
0 & -(1 + i \cdot \cos\gamma_0 \cdot \cos\delta_1) & \begin{array}{l} i \cdot \sin\gamma_0 \cdot \sin\varphi_1 - \\ i \cdot \cos\gamma_0 \cdot \sin\delta_1 \cdot \cos\varphi_1 \end{array} & \begin{array}{l} (-z_{ax} \cdot i \cdot \sin\gamma_0 + a \cdot i \cdot \sin\gamma_0 \cdot \sin\delta_1) \cdot \sin\varphi_1 - \\ \sqrt{p_a^2 + p_t^2} \cdot \sin\delta_1 \cdot \sin\varphi_1 + \\ \sqrt{p_a^2 + p_t^2} \cdot i \cdot \sin\gamma_0 \cdot \cos\delta_1 \cdot \sin\varphi_2 - \\ (-z_{ax} \cdot i \cdot \cos\gamma_0 \cdot \sin\delta_1 + a \cdot i \cdot \cos\gamma_0) \cdot \cos\varphi_1 \end{array} \\[6ex]
1 + i \cdot \cos\gamma_0 \cdot \sin\delta_1 & 0 & \begin{array}{l} i \cdot \sin\gamma_0 \cdot \cos\varphi_1 + \\ i \cdot \cos\gamma_0 \cdot \sin\delta_1 \cdot \sin\varphi_1 \end{array} & \begin{array}{l} (-z_{ax} \cdot i \cdot \cos\gamma_0 \cdot \sin\delta_1 + a \cdot i \cdot \cos\gamma_0) \cdot \sin\varphi_1 + \\ (-z_{ax} \cdot i \cdot \sin\gamma_0 + a \cdot i \cdot \sin\gamma_0 \cdot \sin\delta_1) \cdot \sin\delta_1 - \\ \sqrt{p_a^2 + p_t^2} \cdot \sin\delta_1 \cdot \sin\varphi_1) \cdot \cos\varphi_1 + \\ \sqrt{p_a^2 + p_t^2} \cdot i \cdot \sin\gamma_0 \cdot \cos\delta_1 \cdot \varphi_1 \cdot \cos\varphi_1 - \end{array} \\[6ex]
\begin{array}{l} -i \cdot \sin\gamma_0 \cdot \sin\varphi_1 + \\ i \cdot \cos\gamma_0 \cdot \sin\delta_1 \cdot \cos\varphi_1 \end{array} & \begin{array}{l} -i \cdot \sin\gamma_0 \cdot \cos\varphi_1 - \\ i \cdot \cos\gamma_0 \cdot \sin\delta_1 \cdot \sin\varphi_1 \end{array} & 0 & \begin{array}{l} \sqrt{p_a^2 + p_t^2} \cdot i \cdot \sin\gamma_0 \cdot \sin\delta_1 \cdot \varphi_1 - \\ a \cdot i \cdot \sin\gamma_0 \cdot \cos\delta_1 + \\ \sqrt{p_a^2 + p_t^2} \cdot \cos\delta_1 \end{array} \\[6ex]
0 & 0 & 0 & 0
\end{bmatrix}
$$

$$(3.114)$$

Introducing simplified, symbolic signs for elements of matrix $\underline{P}_{1k}$ as:

$$\underline{P}_{1k}=\begin{bmatrix} 0 & C_1 & C_2 & C_3 \\ -C_1 & 0 & C_4 & C_5 \\ -C_2 & -C_4 & 0 & C_6 \\ 0 & 0 & 0 & 0 \end{bmatrix} \qquad (3.116)$$

For use in computer-aided calculations, these conditions should be kept in mind.

The points of contact curve belong to the φ_1 = const movement parameter or, in other words, the internal parameters u, ϑ determine these points for computer calculations. It is useful to determine the C_i (i = 1, ..., 6) matrix elements and treat them as constants to calculate further points on the contact curve. This method shortens the computational time and makes it possible to follow computations using quick graphs.

Now determine the relative velocity vector using equation (3.110). After completing the operations, we obtain:

$$\left. \begin{aligned} v_{1Fx}^{(12)} &= C_1 \cdot ((u \cdot \cos\beta + p_t \cdot \vartheta) \cdot \cos\vartheta + r \cdot \sin\vartheta) + C_2 \cdot (u \cdot \sin\beta + p_a \cdot \vartheta) + C_3 \\[2mm] v_{1Fy}^{(12)} &= C_1 \cdot ((u \cdot \cos\beta + p_t \cdot \vartheta) \cdot \sin\vartheta - r \cdot \cos\vartheta) + C_4 \cdot (u \cdot \sin\beta + p_a \cdot \vartheta) + C_5 \\[2mm] v_{1Fz}^{(12)} &= C_2 \cdot ((u \cdot \cos\beta + p_t \cdot \vartheta) \cdot \sin\vartheta - r \cdot \cos\vartheta) - C_4 \cdot ((u \cdot \cos\beta + p_t \cdot \vartheta) \cdot \cos\vartheta + r \cdot \sin\vartheta) + C_6 \end{aligned} \right\} \quad (3.117)$$

The normal vector of the surface No1 (worm) can be determined according to equation:

$$\vec{n}_{1F} = \frac{\partial \vec{r}_{1F}}{\partial u} x \frac{\partial \vec{r}_{1F}}{\partial \vartheta} \qquad (3.118)$$

known from differential geometry.
The scalar coordinates of the normal vector are:

$$\left. \begin{aligned} n_{1Fx} &= \frac{\partial y_{1F}}{\partial u} \cdot \frac{\partial z_{1F}}{\partial \vartheta} - \frac{\partial y_{1F}}{\partial \vartheta} \cdot \frac{\partial z_{1F}}{\partial u} \\[2mm] n_{1Fy} &= -\left[\frac{\partial x_{1F}}{\partial u} \cdot \frac{\partial z_{1F}}{\partial \vartheta} - \frac{\partial x_{1F}}{\partial \vartheta} \cdot \frac{\partial z_{1F}}{\partial u} \right] \\[2mm] n_{1Fz} &= \frac{\partial x_{1F}}{\partial u} \cdot \frac{\partial y_{1F}}{\partial \vartheta} - \frac{\partial x_{1F}}{\partial \vartheta} \cdot \frac{\partial y_{1F}}{\partial u} \end{aligned} \right\} \qquad (3.119)$$

After substitutions and completing the operations, the scalar components are:

$$n_{1Fx} = p_a \cdot \cos\beta \cdot \cos\vartheta + \sin\beta \cdot \cos\beta \cdot u \cdot \sin\vartheta - (r + p_t) \cdot \sin\beta \cdot \cos\vartheta + p_t \cdot \sin\beta \cdot \vartheta \cdot \sin\vartheta$$

$$n_{1Fy} = p_a \cdot \cos\beta \cdot \sin\vartheta - \sin\beta \cdot \cos\beta \cdot u \cdot \cos\vartheta - (r + p_t) \cdot \sin\beta \cdot \sin\vartheta - p_t \cdot \sin\beta \cdot \vartheta \cdot \cos\vartheta \tag{3.120}$$

$$n_{1Fz} = \cos^2\beta \cdot u + p_t \cdot \cos\beta \cdot \vartheta$$

Substituting into equation $\bar{n}_{1F} \cdot \bar{v}_{1F}^{(12)} = 0$ expressing the contact conditions (3.117) and (3.120), the equation determining u can express in explicit form the value of internal parameters using the formula for solving the algebraic equation:

$$A_2 \cdot u^2 + A_1 \cdot u + A_0 = 0 \tag{3.121}$$

To solve it numerically the constants (A_2, A_1, A_0) should be determined. The constants after completing the operations and simplifying the expressions are as follows:

$$A_2 = C_2 \cdot \cos\beta \cdot \sin\vartheta - C_4 \cdot \cos\beta \cdot \cos\vartheta$$

$$
\begin{aligned}
A_1 &= ((C_3 + C_4 \cdot p_a) \cdot \sin\beta \cdot \cos\beta - (p_t \cdot \sin^2\beta + r) \cdot C_4) \cdot \sin\vartheta + \\
&+ ((C_2 \cdot p_a - C_5) \cdot \sin\beta \cdot \cos\beta - (p_t \cdot \sin^2\beta + r) \cdot C_2) \cdot \cos\vartheta + \\
&+ (p_a \cdot \sin\beta \cdot \cos\beta + p_t \cdot (1 + \cos^2\beta)) \cdot C_2 \cdot \vartheta \cdot \sin\vartheta - \\
&- (p_a \cdot \sin\beta \cdot \cos\beta + p_t \cdot (1 + \cos^2\beta)) \cdot C_4 \cdot \vartheta \cdot \cos\vartheta + \\
&+ (C_6 + C_1 \cdot p_a) \cdot \cos^2\beta - C_1 \cdot p_t \cdot \sin\beta \cdot \cos\beta
\end{aligned}
$$

$$
\begin{aligned}
A_0 &= (p_a \cdot \cos\beta - (p_t + r) \cdot \sin\beta) \cdot C_5 \cdot \sin\vartheta + \\
&+ (p_a \cdot \cos\beta - (p_t + r) \cdot \sin\beta) \cdot C_3 \cdot \cos\vartheta + \\
&+ (C_3 \cdot p_t \cdot \sin\beta + (p_a^2 \cdot \cos\beta - p_a \cdot p_t \cdot \sin\beta - p_a \cdot r \cdot \sin\beta - p_t \cdot r \cdot \cos\beta) \cdot \\
&\quad \cdot C_4) \cdot \vartheta \cdot \sin\vartheta + ((p_a^2 \cdot \cos\beta - p_a \cdot p_t \cdot \sin\beta - p_a \cdot r \cdot \sin\beta - p_t \cdot r \cdot \cos\beta) \cdot \\
&\quad \cdot C_2 - C_5 \cdot p_t \cdot \sin\beta \cdot \vartheta \cdot \cos\vartheta + (p_a \cdot \sin\beta + p_t \cdot \cos\beta) \cdot p_t \cdot C_2 \cdot \vartheta^2 \cdot \sin\vartheta - \\
&- (p_a \cdot \sin\beta + p_t \cdot \cos\beta) \cdot p_t \cdot C_4 \cdot \vartheta^2 \cdot \cos\vartheta + \\
&+ ((p_a \cdot \cos\beta - p_t \cdot \sin\beta) \cdot p_t \cdot C_1 + C_6 \cdot p_t \cdot \cos\beta) \cdot \vartheta
\end{aligned}
\tag{3.121a}
$$

Using these values, u can be determined numerically as:

$$u_{1,2} = \frac{-A_1 \pm \sqrt{A_1^2 - 4 \cdot A_2 \cdot A_0}}{2A_2} \tag{3.122}$$

To obtain the numerical values for u, ϑ belonging to a given point of the contact curve, use ϑ = constant to select from the two roots the possible one to meet practical requirements.

Out of the possible values of u, those which determine the value of y_2 within the contact domain of worm and grinding wheel (r_{f1} − r_{a1}) should be chosen.

Using the equations:

$$\vec{r}_{1F} = \vec{r}_{1F}(u_0, \vartheta_0) \tag{3.123}$$

and

$$\vec{r}_{2F} = \underline{M}_{2F,1F} \cdot \vec{r}_{1F}(u_0, \vartheta_0)$$

the points belonging to the given φ_1 and φ values can be determined in the coordinate system joined to element No2. The points of the contact curve can be determined arbitrarily by giving values within the operational domain.

To review the steps of generation profile of tool in general form, the equation system describing the contact curve when φ_1 = const movement parameter is:

$$\left. \begin{array}{l} \vec{n}_{1F} \cdot \vec{v}_{1F}^{(12)} = \vec{n}_{1F}(u, \vartheta) \cdot \vec{v}_{1F}^{(12)}(u, \vartheta) = 0 \\ \vec{r}_{1F} = \vec{r}_{1F}(u, \vartheta) \\ \varphi_1 = \text{const} \end{array} \right\} \tag{3.124}$$

Solving the system of equations, the contact curve determined for the φ internal parameter is:

$$\vec{r}_{1F} = \vec{r}_{1F}(u(\vartheta), \vartheta) = \vec{r}_{1F}(\vartheta) \tag{3.125}$$

Utilizing the equation of transformation:

$$\vec{r}_{2F}(\vartheta) = \underline{M}_{2F,1F} \cdot \vec{r}_{1F}(\vartheta) = \begin{bmatrix} x_{2F}(\vartheta) \\ y_{2F}(\vartheta) \\ z_{2F}(\vartheta) \\ 1 \end{bmatrix} \tag{3.126}$$

the $R_k = R_k(z_{2F})$ function of the tool profile can be obtained by substituting ϑ into the first equation:

$$\left. \begin{array}{l} R_k = \sqrt{x_{2F}^2(\vartheta) + y_{2F}^2(\vartheta)} \\ z_{2F} = z_{2F}(\vartheta) \end{array} \right\} \tag{3.127}$$

This procedure can be applied when the surface No2 is a block of rotation with axis z_{2F}.

3.2.2.4 Optimization of cutting tool profile

In this section we investigate the generation of the optimized tool profile in the case of a general conical helicoid.

In machining practice the tool surface on element No2 uses its surface of revolution or, with regular edge geometry, the enveloping surface of its machining edges (eg circular milling cutter). The function of the tool profile in such cases was considered in the previous chapter.

To consider the surface element No2 generated as the contact surface of the general conical helicoid described in the previous chapter, after substituting the A_2, A_1, A_o, C_i (i = 1, ..., 6) coefficients into the formulae to calculate the internal parameter u, it can be seen that its value depends on gearing ratio i_{21}. This means that the surface of element No2 is not a surface of revolution because it cannot be rotated itself, so it cannot be a tool surface in practice. Dividing by i_{21}, both numerator and denominator i_{21} remains only in the numerator.

The fractions, being in the numerator, contain the gearing ratio i_{21} only in their own denominator. This means that by increasing gearing ratio i_{21}, the effect on changes in the contact line is decreasing. So in practice with the gearing ratio used for grinding, the effects of elements causing changes can be neglected, as is verified by the calculations. As a result, the points of tool profile function that can be obtained by rotating the points of contact line into the oblique plane fits on axis z_{2F}.

The task is more complicated; the exact solution of it requires continuous control as the tool-profile function shape is a function of the movement parameter φ_1: the acting thread profile, situated between the conical dedendum and addendum surfaces, is changing along its length of the tool profile. This change of general conical thread surfaces is brought about by two factors:

■ We have already noted that when machining Archimedian conical thread surfaces at points belonging to the same value of the internal parameter u, the curvatures characterizing the surface are changing owing to the change of radius with the value of $p_t \times \vartheta$.

■ Besides this, another effect causes changes in tool profile during the machining of involute and convolute conical surfaces. The origin of this effect is that points belonging to the same value of internal parameter u depart from each other in accordance with a hyperbolic function instead of the linear function determined by kinematic relations. This hyperbolic function approaches the linear path realized by kinematic relations. Increasing the diameter of either the base or the dedendum cylinder, the difference between the paths increases.

In spite of all of these factors, the machining of conical helicoid surfaces can be carried out within allowable technical limits, using a previously controlled profile grinding wheel. When the realized surfaces deviation differs from the theoretical one, it is useful to determine the optimal value of the movement parameter $\varphi_{1opt.}$, defined as the value that guarantees the minimum value of inclination. This optimal value, in practice φ_{1op}, gives the average tool profile

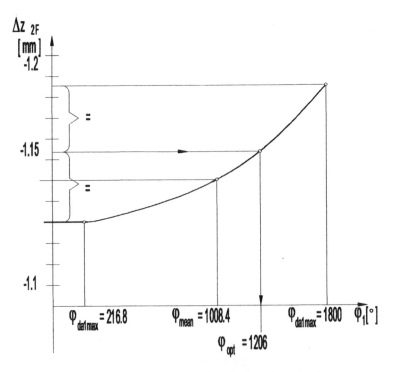

Figure 3.31 *Determination of the optimal place ($\varphi_{1opt.}$) of grinding wheel adjustment*

between the maximum and minimum diameters of the conical helicoid surface. As the change in tool profile as a function of φ_1 is not linear, the optimal value does not belong to the mean diameter of the helicoid surface (see Figure 3.31).

The optimization can be carried out using a graphical method, fulfilling the requirements of precision in practice. Presented in the diagram are the calculated values as a function of φ_1 (see Figure 3.31). Another way of optimization uses an iterative method using the software described in Figure 3.32, and its results are shown in Figure 4.17.

The grinding wheel profile for machining a conical worm is shown for several diameters of the worm in Figure 3.33. The figure represents the correlation between the optimized tool profile and worm profile in axial section.

Geometrically optimum manufacture of conical helicoid surfaces using continuous dressing of the grinding wheel

The investigations carried out have made it clear that the exact machining of conical helicoid surfaces using existing methods provide approximately precise results only.

It is clear that geometrically precise machining can only be realized by changing the profile of the tool during the machining. This seemingly impossible condition can be realized using a CNC-controlled grinding wheel regulator to change the wheel profile, as its momentary position requires (Dudás, 1988a) (Figures 5.20 and 7.14). This regulating movement can be built up from frequently alternating, profile-regulating movements, producing on the wheel surface 'steps' within the range surface roughness of the helicoid surface. Using this method the default of the previously used optimizing method can be reduced in stages by increasing the number of controlled movements.

Investigating the change of grinding wheel profile, the following conclusions may be drawn:

■ The changes of profile along the length of worm are unequal, that is, if its profile defined by the mean diameter of the worm when machining is applied to a grinding wheel, the value of profile deviation will differ at the two ends of the worm;

■ The remaining geometric deviation can be reduced even when finishing with a grinding wheel having a profile determined by the mean diameter of worm, giving an additional (or secondary)

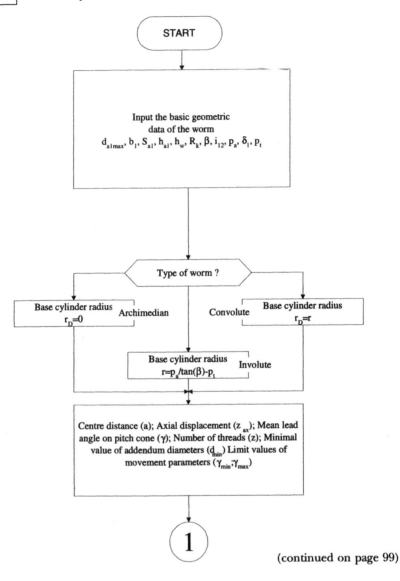

(continued on page 99)

- Addendum diameter of worm d_{a1max}
- Acting length of worm b_1
- Thickness on tip of worm tooth s_{a1}
- Addendum height of worm tooth h_{a1}
- Active tooth height h_w
- Radius of grinding wheel r_k

- Tooth profile angle β
- Gearing ratio i_{21}
- Axial parameter of thread p_a
- Half cone angle of the worm addendum surface d_1
- Tangential parameter of thread p_t

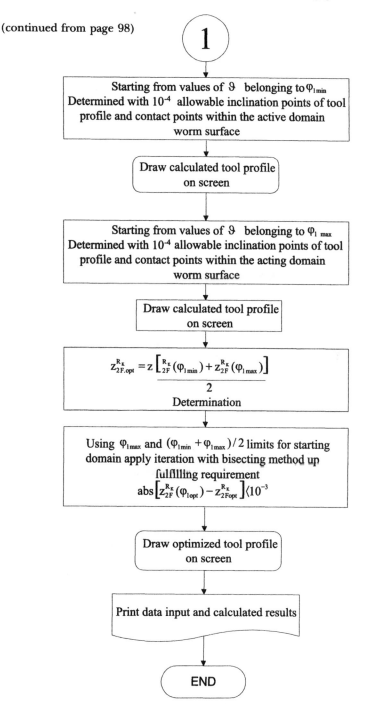

(continued from page 98)

1

Starting from values of ϑ belonging to φ_{1min}
Determined with 10^{-4} allowable inclination points of tool
profile and contact points within the active domain
worm surface

Draw calculated tool profile
on screen

Starting from values of ϑ belonging to $\varphi_{1\,max}$
Determined with 10^{-4} allowable inclination points of tool
profile and contact points within the acting domain
worm surface

Draw calculated tool profile
on screen

$$z_{2F.opt}^{R_K} = z \frac{\left[z_{2F}^{R_K}(\varphi_{1min}) + z_{2F}^{R_K}(\varphi_{1max})\right]}{2}$$

Determination

Using φ_{1max} and $(\varphi_{1min} + \varphi_{1max})/2$ limits for starting
domain apply iteration with bisecting method up
fulfilling requirement
$$\mathrm{abs}\left[z_{2F}^{R_K}(\varphi_{1opt}) - z_{2Fopt}^{R_K}\right] \langle 10^{-3}$$

Draw optimized tool profile
on screen

Print data input and calculated results

END

Figure 3.32 *Determination of optimal tool profile*

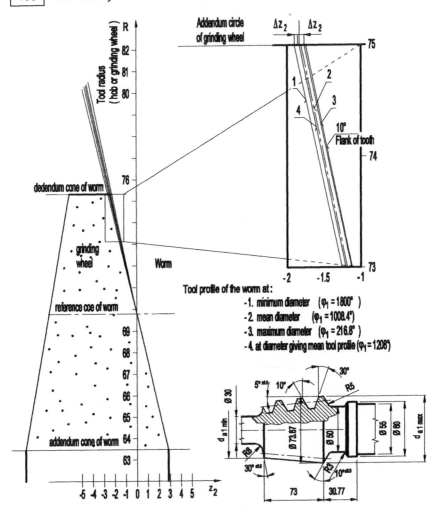

Figure 3.33 *Profile in axial section of grinding wheel for spiroid worm-generating optimal geometry on conical helicoid surface (a = 100 mm, m = 5 mm, i = 41, $z_1$1, β_j = 10°, β_6 = 30°)*

motion to the grinding wheel, producing changes in deviation between the axes of the tool and worm during machining. Generalizing this consequence, a machine tool for grinding is required that it is possible to regulate by rotating the direction of the grinding wheel shaft round two axes parallel to the machining. In this way, the relative position of the grinding wheel and worm is continually changing so that the deviation between the contact curves forms at a given time. At the same time it is possible to ensure that the grinding wheel profile remains constant during machining so that, applying the wheel movements mentioned above, the deviation in geometric shape is minimal compared to wheel shape as well;

- The third conclusion concerning the grinding of thread surfaces is that the geometric errors mentioned before would not occur when the profile of the grinding wheel changes during machining. The fulfilment of this condition seems impossible at first, but a traditional wheel reprofiled using diamond, and reprofiling parallel to the grinding, should be applied coordinated with displacement of the grinding wheel along the thread surface. When the allowable step formed on the wheel is 0.001 mm in thickness then a velocity of about 500 mm/minute for regulating the diamond is required.

The pre-finishing grinding can be carried out with a wheel profile of the minor diameter of the worm, profiling it before grinding. If the profile difference between the two ends of the worm is small or the profiling can be carried out in parallel with the grinding, then the wheel should be prepared as a major diameter of the workpiece required left on oversize because, owing to the possible profile error, this can only be removed using multiple grinding cycles that always start at the major diameter of worm. The last, finishing sparkling circle of grinding parallel to profiling should be started at the major diameter of the profile and completed at the profile belonging to the minor diameter of worm. A ground conical spiroid worm can be seen in Figure 8.23.

The problem is soluble if, during grinding, the wheel profile requires additional material that had been removed by regulating it (eg globoid threads). This time an additional feed of tool should be added comparable with workpiece displacement.

This is a brand new machining technology with the tool changing its profile during the manufacturing procedure.

Summarizing, the innovation discussed in this chapter is the new technology, with the tool changing its profile during machining because, in the theory of manufacturing technology up to now, the rigid tool was regarded as the foundation stone.

Research work carried out by the author's team has proved that the exact solution of problems during machining or more precisely machining problems involving a special circle of workpiece requires a new technology that can be realized with the help of NC- control-led machine tools. In this way the unified thread grinding machine and wheel profiling device coordinated with parallel co-working is guaranteed, while the grinding wheel performs parallel roles of the thread surface generating tool and the reprofiling diamond while manufacturing the perfect shape of the element.

The advantage of this method is that the profile difference be-tween the limiting positions is small. The device and procedure suitable for the realization of this principle will be discussed in Chapter 5.

3.3 GEOMETRIC ANALYSIS OF HOBS FOR MANUFACTURING WORM GEARS AND FACE-GEARS MATED CYLINDRICAL OR CONICAL WORMS

From the point of view of machining technology, the fly cutter can be regarded as one tooth of a worm hob tool so the surfaces to be manufactured can be considered to be equivalent. When manufac-turing fly cutters and worm hobs the difference is only in their domains of movement and their realization in practice. The border surfaces and cutting edge of a single tooth of the tool are illustrated in Figure 3.34.

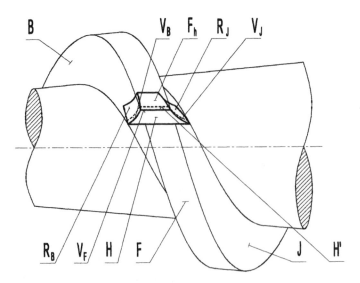

Figure 3.34 *Surfaces of worm hob*

The most important surfaces, borders of the tool, in Figure 3.34 are:

- H — face surface;
- R_B, R_J — back surfaces (left and right);
- F_h — back surface of head ribbon;
- B, J — side surfaces of basic hob worm.

The cutting edges are:

- V_B — left side edge as intersection of B tooth and H face surfaces;
- V_J — right side edge as intersection of J tooth and H face surfaces;
- V_F — addendum edge as intersection of F head ribbon and H face surfaces.

The author's investigations are first concerned with edges V_B and V_J as basic influencing factors in the precision of a worm gear as well as back surfaces R_B and R_J and face surface H, because these surfaces determine the edges investigated.

Let $\vec{r}_g$ (ξ, η, ζ) be the curve of the B helicoidal surface having parameters p_a (axial) and p_r (radial thread) in the form:

$$\vec{r}_{1F}{}^{B}(\eta, \vartheta) = \underline{M}_{1F,0} \cdot \vec{r}_g \qquad (3.128)$$

To face surface H is a ruled (closed and flat) helicoidal surface with a generator line perpendicular to the axis of surface B showing a lead direction perpendicular to the lead direction measured on the pitch cone of helicoid B. This surface can be defined as:

$$\vec{r}_{1F}{}^{H} = \vec{r}_{1F}{}^{H}(\eta, \vartheta) \qquad (3.129)$$

The suffix 1F referred to as the hob tool in the author's investigations is regarded as the workpiece to be manufactured and in accordance with the symbols used for machining geometry the surfaces are defined in the K_{1F} coordinate system rotating with the workpiece.

The obtain the parametric equation of the cutting edge V_B, eliminate either parameter η or ϑ from equations (3.128) or (3.129); that is:

$$\vec{r}_{1F}{}^{VB} = \vec{r}_{1F}{}^{VB}(\eta) \qquad \text{or} \qquad \vec{r}_{1F}{}^{VB} = \vec{r}_{1F}{}^{VB}(\vartheta) \qquad (3.130)$$

While the characteristics of side edges V_B, V_J have a role in shaping up the machined surface, the face surface H has a significant

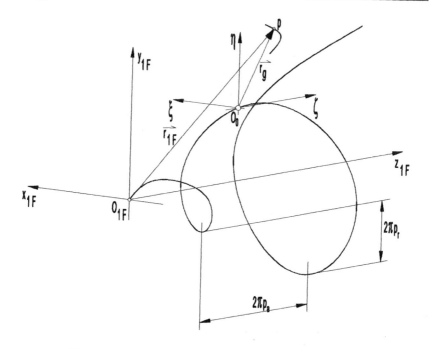

Figure 3.35 *Generation of conical helicoid surface*

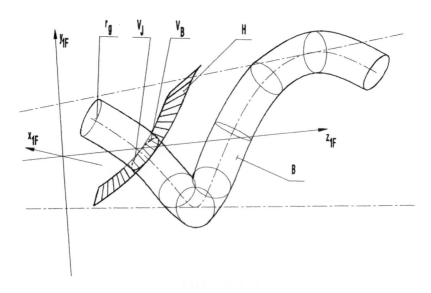

Figure 3.36 *Cutting edge V_B generation*

effect on the service life of the tool and on chip generation. The shapes of the back surfaces R_B, R_J influence the process of machining and the possibilities for resharpening the tool. As surface-generating hobs are expensive tools having a complicated geometric shape it is important to make resharpening possible several times, something to be taken into consideration when designing them.

Resharpening also generates a new tooth surface. That is, this process can be considered to be the choosing of a suitable tooth surface H, realized in the kinematic conditions of the resharpening device, and moving it so that a series is generated while the edge V_B (or V_J), depending on the path followed by back surface R_B (or R_J), will be generated (see Figure 4.42).

To keep the shape of the V_B (V_J) with the cutting edge unchanged during several resharpenings, the technology of resharpening, shaping the new H' face surface, should be chosen by taking into consideration the generation of the back surface R_B (R_J) too.

It should be mentioned that keeping the original shapes of cutting edges V_B, V_J while resharpening is not satisfactory in itself to guarantee that the generated teething on the workpiece be preserved as unchanged.

Now we shall investigate practical ways to generate back surfaces (see Figure 3.37). Out of these, possible ways will be to choose the minimum value of back angle that results in optimal conditions during machining consistent with simple manufacture.

The mathematical description of the tool back surface is analogous to the formulation of a conical helicoidal surface. The role of the generator curve r_g is taken over by the side edge V_B cited as a vector–scalar function for the intersection of the worm B and face surfaces H. The movement made by the relief machine tool is determined by the transformation matrix differing from the helicoidal surface generating matrix in the numerical values of thread parameters p_a and/or p.

- Relief is described as *axial* when the axial thread parameter p_a^R of the hob back surface differs from the axial thread parameter p_a of the helicoid surface, but radial thread parameter p_a^R is equal to p_r, the radial thread parameter of the helicoid surface.
- Relief is described as *radial* when the radial thread parameter p_a^R of back surface R_B differs from radial thread parameter p_r of helicoid surface B but values p_a^R and p_a are equal.

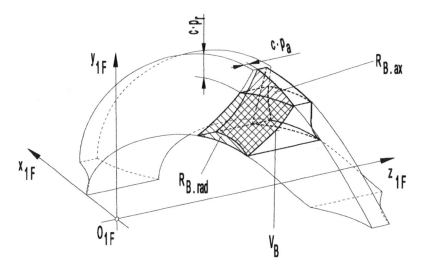

Figure 3.37 *Axial and radial relief*

- *Diagonal relief* occurs when thread parameters are equal, that is $p_a^R \neq p_a$ and $p_a^R \neq p_r$. It goes without saying that should it be chosen out of smaller or bigger relations ($p_a^R \neq p_a$ and $p_a^R \neq p_r$), the value results in the back surface facing into the block.

We shall now consider the effect of tooth shape of the generating hob on the geometric precision of the teething generated, in relation to its resharpening.

The factors to be considered are:

- number of edges;
- axial thread parameter of back surface p_a^R;
- radial thread parameter of back surface p_a^R.

The effects of the above three parameters will be investigated separately for:

- cylindrical meshing surface B ($p_r=0$);
- conical meshing surface B ($p_r>0$).

Case of cylindrical meshing surface B; $p_r = 0$

- *Number of edges:* The number of edges has no effect on the geometric precision of the worm gear (neglecting wear of cutting edge) so, in practice, the single cutting edge tool, the fly cutter, can be used for tangential machining. In the case of radial machining, precision of the worm gear can be increased by increasing the number of cutting edges (Litvin and Feng, 1996).

- *Axial thread parameter of back surface* p_r^R : The edge obtained by resharpening (V_B or V_J) with axial relief by axial readjustment of the tool parallel to its rotation around its axis can be fitted in the position of the original cutting edge. That is why it is suitable for generating a geometrically precise tooth surface on the worm gear. As a consequence of resharpening, the distance between cutting edges V_B and V_J decreases. This can be prevented using the method, developed by the Klingelnbergy company, who adjusted axially their two-piece hob by distance ring fixing. The number of possible resharpenings is limited by the tooth height reduction apparent in a decrease in head clearance between mated elements. For guaranteed suitable head clearance the total length of the cutting tooth can be utilized for resharpening. Naturally, the minimum length required from the point of view of required strength should be taken into consideration.

- *Radial thread parameter of back surface* p_r^R : The edge obtained by resharpening (V_B or V_J) in the case of radial relief ($p_r^R \neq 0$) approaches the axis of the tool while the distance between left and right edges remains constant. A wrapping helicoid surface over the new cutting edges can be obtained from the original helicoid surface B by decreasing it radially, so a new surface is shaped up, but the worm gear manufactured by the new cutting edges will be a defective one. Increasing the thread parameter causes dimensional deviation. The differences in dimensions can within limits be reduced by reducing the distance between the axes of the tooth-cutting tool and the worm gear as well as by readjusting the lead angle difference. However, taking into account the narrow tolerances of worm gear the number of possible resharpenings is small (Grüss, 1951). When radial 're-lief turning' is mated with axial 'relief turning', that is, when diagonal 'relief turning' is applied, it can be supported by highly curved edges, and the geometric deviations are curtailed.

Case of conical wrapping surface $p_r > 0$

- **Number of edges:** Before considering the number of edges, it should be noted that enveloping surface B, having a uniform lead on a conical base, is itself non-movable. At the same time the elements having conical active surfaces can in practice be twisted from their own position (the axial profile has no part inclining backwards (Figure 3.38)).

Because of this, the conical generating hob can, by axial displacement, reach into contact position with the worm without harm to the active surface on the teethed worm gear.

This still does not guarantee that the edges of the tool moved into the worm and rotated there mesh with the worm gear surface equivalent to enveloping surface B or that the mated worm would result.

This results in the fact that a limited number of cutting edges only can be shaped over the conical hob. It is easy to understand that by decreasing the number of cutting edges, the precision of the toothed surface is made worse. In an extreme case, when leaving a single cutting edge at the minor diameter, as in a blade cutter, the fault becomes evident. The geometric deviations of a toothed wheel can be reduced by increasing the number of teeth on the generating hob (in cylindrical worms it is known as a *hulling worm*) or by applying a series of generating hobs having several teeth with different starting positions over the generating hob surface B.

The precision of the wheel can be increased by increasing the number of teeth on the hob when machining with radial displacement.

■ *Axial thread parameter of back surface* p_a^R : If the cutting edges are spaced apart by equal angular displacement θ and radial relief $p_a^R \neq p_r$ applied to the edges of the resharpened hob, they can be fitted on the original conical helicoid surface B. This is illustrated in Figure 3.38.

As the relative position of edges V_B and V_J in axial direction remains unchanged on resharpening, owing to the radial relief, the best direction for relief hobs with conical wrapping surface B is the radial one.

During design, the length of the generating hobs should be taken into consideration with the axial displacement $\theta.p_a$ belonging to central angle θ representing the maximal angle for resharpening. At the same time, the axial position of the tool should be readjusted by the value calculated using the central angle θ' belonging to the actual resharpening of the tool.

Machining axial direction of back surfaces

In the machining of axially 'relieved' back surfaces, with the periodicity of hob teeth and the repeated machining of each blade

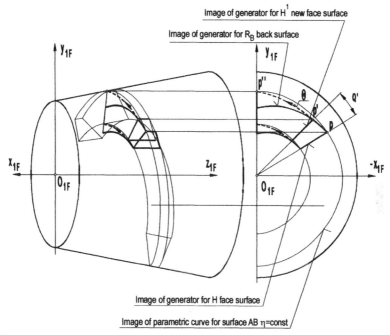

Figure 3.38 *Rotation of new edge point P' on conical helicoid surface B*

surface, new problems arise concerning machining geometry. The kinematics of machining and the geometry of the generating tool can be determined as discussed in section 3.1. The head ribbon should be relieved in the radial direction.

Machining radial direction of back surface

Using the kinematic model and tool surface-determining algorithm for finishing conical helicoid surfaces, as reviewed in sections 3.2.2.3 and 3.2.2.4, the machining of back surfaces and surfaces for grinding tools can be determined. To avoid collision between the tools (grinding wheel during the relief and the worked tooth surface) a pinned or hollow cylindrical shape grinding wheel can be used where the coordinate system will be chosen as the tool adjustment requires it (Figure 4.40).

For machining of a back surface generated by radial relief, the best solution is the CNC machine tool mated with a continuous wheel control device. When no continuous control is applied, the wheel can be optimized in the domain of conical surfaces as partial surfaces of the back surface for all the teeth on the milling cutter using one of the procedures described in section 3.2.5.

3.3.1 Investigation of cutting tool for manufacturing worm gear mated with worm having arched profile

The shape of a worm-generating tool and its position during machining determine the geometry of the worm teething, and vice versa, the geometry of the worm determines the tool used for machining the worm gear. The consequence of this is that the teething of a tool-generating worm can be a generally used one, while the teething of worm gear should be machined using individual, special tools.

The working surfaces of worm and of worm gear mutually mesh with each other; for this reason, when finishing worm gear, a tool should be used having the equivalent geometry of the worm.

For individual and small-scale manufacture, the fly cutter is used while for mass production the generating worm hob is used when machining worm gear.

When the teeth ratio z_2/z_1 is not a whole number, the machining of worm gear by flyblades is easier and reduces greatly the effect of pitch errors on smooth operation.

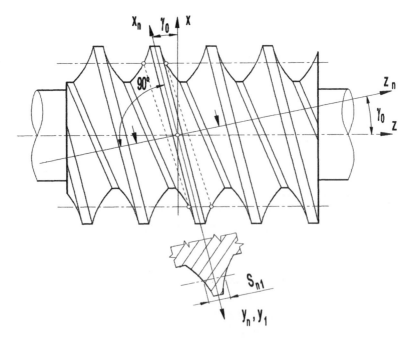

Figure 3.39 *Defining the normal plane*

When the teeth ratio z_2/z_1 is a whole number, this makes possible the running in of all the flanks of the teeth with pitch errors of the worm; but, for manufacturing, the use of the generator milling cutter for worm gear is inevitable.

Because the worm is ground, the tool used for worm gear machining (fly cutter or worm milling cutter) can be prepared in a direct way, that is the machined worm equivalent (see Figures 3.45 and 3.46).

In this way the profile of the fly cutter is equivalent to the profile of the worm in normal section and the generated tooth surfaces will be the conjugates of each other. Keep in mind that grinding of worm and fly cutter (machining of land) can be done by axial displacement (applying from side).

3.3.1.1 Equation of hob profile for manufacturing worm gear

The tooth face plane of the fly cutter or milling cutter is equivalent to the normal section of the worm. That is, the equation of its profile can be obtained by intersecting the worm parametric equation with the normal plane. The normal plane is defined in stationary coordinate systems (Figure 3.39).

Equation of the normal sectioning plane is:

$$x_1 = z_1 \cdot \text{tg} \gamma_0 \qquad (3.131)$$

and the parametric equation system of the worm:

$$
\left.
\begin{aligned}
x_1 &= -\eta \cdot \sin \Theta \\
y_1 &= \eta \cdot \cos \Theta \\
z_1 &= p \cdot \Theta - \sqrt{\rho_{ax}^2 - (K - \eta)^2} + z_{ax}
\end{aligned}
\right\}
\qquad
\begin{aligned}
&(3.132/1) \\
&(3.132/2) \\
&(3.132/3)
\end{aligned}
$$

The simultaneous solution determines the equation of the tool edge as:

$$
\left.
\begin{aligned}
y_n &= \eta_n \cdot \cos \Theta \\
z_n &= -\frac{\eta_n \cdot \sin \Theta}{\sin \gamma_o}
\end{aligned}
\right\}
\qquad (3.133)
$$

The variable η_n can be obtained by solving equations (3.131) and (3.132).

Using equation (3.132/2) we obtain:

$$\eta_n = \frac{y_1}{\cos \Theta}$$

and from equations (3.131) and (3.132/1):

$$z_1 \cdot tg\gamma_0 = -\eta \cdot \sin\Theta$$

$$z_1 = \frac{-\eta_n \cdot \sin\Theta}{tg\gamma_0} = \frac{-y_1 \cdot tg\Theta}{tg\gamma_0}$$

Substituting z_1 and h_n into equation (3.132/3):

$$-y_1 \cdot \frac{tg\Theta}{tg\gamma_0} = P \cdot \Theta - \sqrt{\rho_{ax}^2 - (K - \eta)^2} + z_{ax} \qquad (3.134)$$

After reorganizing and completing the assigned operations the result is:

$$y_1^2 \cdot (\sin^2\Theta + tg^2\gamma_0) \cdot \frac{1}{\cos^2\Theta} - 2 \cdot y_1 \cdot \left[\sin\Theta \cdot tg\gamma_0 \cdot (P \cdot \Theta + z_{ax}) + tg^2\gamma_0 \cdot K\right] \cdot \frac{1}{\cos\Theta} +$$
$$+ tg^2\gamma_0 \cdot \left[(P \cdot \Theta + z_{ax})^2 + (K^2 - \rho_{ax}^2)\right] = 0$$

$$y_1^2 \cdot A \cdot \frac{1}{\cos^2\Theta} - 2 \cdot y_1 \cdot B \cdot \frac{1}{\cos\Theta} + C \cdot tg^2\gamma_0 = 0$$

Substituting $\eta_n = \frac{y_1}{\cos\Theta}$ the equation can be obtained as:

$$A \cdot \eta_n^2 - 2 \cdot B \cdot \eta_n + C = 0 \qquad (3.135)$$

Solving for variable η_n

$$\eta_n = \frac{2 \cdot B \pm \sqrt{4 \cdot B^2 - 4 \cdot A \cdot C}}{2 \cdot A} = \frac{B}{A} \cdot \left[1 \pm \sqrt{1 - \frac{A \cdot C}{B^2}}\right] \qquad (3.136)$$

where

$$\left.\begin{array}{l} A = \sin^2\Theta + tg^2\gamma_0 \\ B = \sin\Theta \cdot tg\gamma_0 \cdot (p \cdot \Theta + z_{ax}) + K \cdot tg^2\gamma_0 \\ C = tg^2\gamma_0 \cdot \left[(p \cdot \Theta + z_{ax})^2 + (K^2 - \rho_{ax}^2)\right] \end{array}\right\} \qquad (3.137)$$

When calculating z_{ax}, the tooth clearance j_n necessary for continuous operation of the worm gear drive should be taken into consideration. As the coordinate axis y_1 is situated in the symmetry plane of the tooth:

$$z_{ax} = \frac{\overline{S_{nt}}}{2} + \frac{j_n}{2} + \sqrt{\rho_{ax}^2 - (K - \eta)^2} = \frac{\overline{S_{sz}}}{2} + \sqrt{\rho_{ax}^2 - (K - \eta)^2} \qquad (3.138)$$

Carrying out direct numerical calculations with the derived equations, it can be concluded that values obtained are equal to the values measured on the profile of the manufactured fly cutter.

The coordinates of the tool edge obtained from the solution of (3.135) and (3.137) can be used both for machining and checking (preparing 'go' or 'no go' gauges).

3.3.1.2 Shape of the conventional fly cutter

The fly cutter is ground on the same machine as the worm so that the pre-manufactured fly cutter is fitted in place of the worm to grind the side surface of the previously hardened and pre-finished edge ribbon fly cutter. After the grinding of the side surfaces a few tenths of 1 mm of land should be kept on. Resharpening of the blade is carried out on its face without removing it from the case (see Figure 3.47). The blade after resharpening forms the same tooth profile up to the total using up of the land.

Applying straight relief, this is applied most frequently in practice; that is, no land is left. Then a theoretically precise worm gear profile can be obtained only until first resharpening.

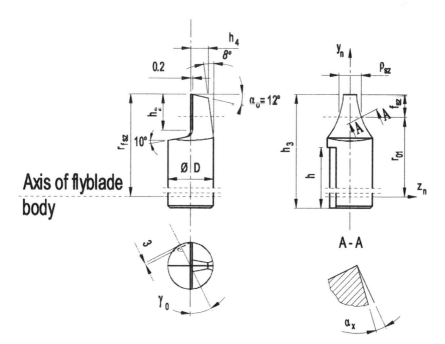

Figure 3.40 _Traditional shape of fly cutter_

The traditional fly cutter is presented in Figure 3.40. Its clamping arrangement is shown in Figure 3.45.

3.3.1.3 Geometry and calculation of hob for cylindrical worm gear or flyblade with helicoidal face surface

The teething of a worm gear is determined by the teething of the worm with contact surfaces mutually meshing with each other. At the finishing, ie the teething of the worm, such a tool should be applied so that its cutting edges are situated on the surface of a substituting worm.

This substituting worm is similar to the real worm gear with which it will be mated. They have common geometric axes, reference cylinders are equal and a single side of their teeth coincides when applying a suitable axial displacement. The addendum diameter and tooth thickness of the substituting worm is larger than the real one (Figure 3.41).

During machining, when the substituting worm and worm gear are mated, the centre distance is that required by the ready-made worm gear and its own conjugated worm.

At the beginning of section 3.3 it was mentioned that the cutting edges of a hob for finishing worm gear should be situated on the surface of the substituting worm, the equivalent of the real worm, which will be mated with the ready-made worm gear (see Figure 3.41).

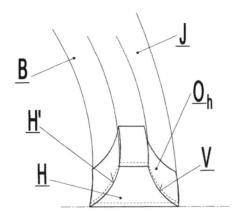

Figure 3.41a *Edges of hob and surfaces of substituting worm*

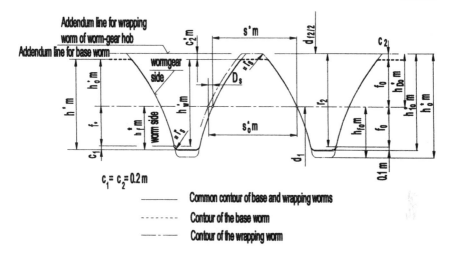

Figure 3.41b *Base profiles of worm, worm gear and generating milling cutter*

The thickness of the threads on the worm substituting the hob and the diameters of the addendum and dedendum cylinders differ from the corresponding sizes of the real worm. The reference cylinder diameter da of the finishing worm gear hob is equal to the reference cylinder diameter of the worm.

As the side cutting edge is the intersection (V) of the relieved side surface (R_b, R_j) and face surface (H), both the relief and the machining of face surfaces should be carried out so that the generated edge fits tooth surface au of the substituting worm (J, B), the equivalent of the tooth surface of the real worm.

The cutting edge V can be obtained as an intersection of the three surfaces:

- the relief side surface R, and/or R_b;
- the face surface H;
- tooth surface of substituting worm J and/or B.

The cutting edge V should remain on the surface of the same worm after resharpening; axial displacement is allowed. In this way the tool, after its resharpening, generates a deformation-free tooth surface, the only difference in tooth thickness that can occur being on the machined worm gear.

To fulfil previously discussed requirements both the relieved side surface and the face surface should be constant lead cylindrical helicoid surfaces.

This is the only case in which the relative positions of intersection line and worm shaft remain constant if one of the tooth surfaces is subjected to angular displacement, for example, if owing to resharpening, the H face surface rotates into the H surface. As a consequence, for a face surface a helicoidal surface is chosen; in this case it is a (closed, flat) helicoid surface. The angle between the thread line on the reference cylinder of this surface and its centre line on the reference cylinder of the surface and its centre line equals γ_0, the lead angle of the thread line on the reference cylinder. In some cases this surface can be a plane parallel to the axis or placed on the axis. It is appropriate to place a plane on the worm axis only if $\gamma_0 \leq 3\text{--}5°$ because in the case of higher lead angles, the machining conditions are not satisfactory.

As the relief surface is generated by the intersection of the tooth side surface of the substituting worm and the face surface of the gash, this must first be determined.

Determination of thread-face surface on the hob

The face surface of the hob is an appropriate closed Archimedian thread surface along which the resharpening of the tool may be carried out.

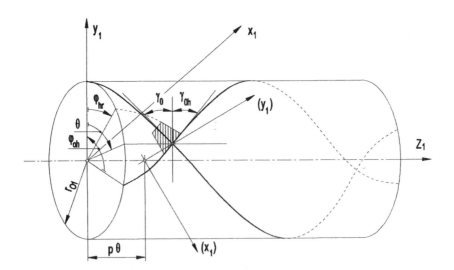

Figure 3.42 *Generation of thread face and its relation with back surface*

Generation of this surface: a half-line perpendicular to the centre line of the helicoid is drawn in accordance with face thread parameter p_h following a translational motion parallel to its rotation, too.

On a right-hand thread-milling cutter, its face surface follows a left-hand thread, where the equation of the face surface based on Figure 3.42 is:

$$\left. \begin{array}{l} x_h = -\eta \cdot \sin(\vartheta + \varphi_{oh}) \\ y_h = +\eta \cdot \cos(\vartheta + \varphi_{oh}) \\ z_h = -p_h \cdot \sin(\vartheta + \varphi_{oh}) \end{array} \right\} \qquad (3.139)$$

and the lead parameter of the face surface is:

$$p_h = \frac{d_{01}\pi \cdot \tan\gamma_{oh}}{2 \cdot \pi}$$

The intersection of the surface (3.12) and the face surface (3.139) is determined by the condition $z_h = z_1$.

Out of this equation:

$$\varphi_{oh} = \frac{1}{p_h} \cdot \sqrt{p_{ax}^2 - (K - \eta)^2} - \frac{p}{p_h} \cdot (\vartheta + \varphi_0) - \vartheta - \frac{z_{ax}}{p_h} \qquad (3.140)$$

which after substituting it into (3.139) can be written:

$$\left. \begin{array}{l} x_h = -\eta \cdot \sin\left[\frac{p}{-p_h} \cdot (\vartheta + \varphi_0) + \frac{1}{p_h}\sqrt{p_{ax}^2 - (K - \eta)^2} - \frac{z_{ax}}{p_h} \right] \\ y_h = \eta \cdot \cos\left[\frac{p}{-p_h} \cdot (\vartheta + \varphi_0) + \frac{1}{p_h}\sqrt{p_{ax}^2 - (K - \eta)^2} - \frac{z_{ax}}{p_h} \right] \\ z_h = p \cdot (\vartheta + \varphi_0) - \sqrt{p_{ax}^2 - (K - \eta)^2} + \frac{z_{ax}}{p_h} \end{array} \right\} \qquad (3.141)$$

As $y = y_h$:

$$\eta \cdot \cos(\vartheta + \varphi_0) = \eta \cdot \cos\frac{p \cdot (\vartheta + \varphi_0) - \sqrt{p_{ax}^2 - (K - \eta)^2} + z_{ax}}{-p_h}$$

and:

$$\Theta = \vartheta + \varphi_0$$

then:

$$\Theta = \frac{\sqrt{p_{ax}^2 - (K - \eta)^2} + z_{ax}}{p + p_h}$$

By substitution, the equation of the cutting edge becomes

$$
\left.
\begin{aligned}
x_v &= -\eta \cdot \sin \frac{\sqrt{\rho_{ax}^2 - (K-\eta)^2} - z_{ax}}{p+p_h} \\[2mm]
y_v &= \eta \cdot \cos \frac{\sqrt{\rho_{ax}^2 - (K-\eta)^2} - z_{ax}}{p+p_h} \\[2mm]
z_v &= -p_h \cdot \frac{\sqrt{\rho_{ax}^2 - (K-\eta)^2} - z_{ax}}{p+p_h}
\end{aligned}
\right\}
\tag{3.142}
$$

Equations of the relief side surfaces

Let the parameter of the relief thread surface be p'; then the equation of the back surface using the equation of the worm surface (3.12) is:

$$
\left.
\begin{aligned}
x_{hr} &= -\eta \cdot \sin(\vartheta + \varphi_{hr}) \\
y_{hr} &= \eta \cdot \cos(\vartheta + \varphi_{hr}) \\
z_{hr} &= p'(\vartheta + \varphi_{hr}) - \sqrt{\rho_{ax}^2 - (K - \eta^2} + z_{ax}
\end{aligned}
\right\}
\tag{3.143}
$$

where:

p' = p' (h$_r$);
y' = y' (h$_r$);
h$_r$ value of relief;
φ_{hr} as determined by Figure 3.42.

The intersection of the relief side surface (3.139) can be determined when $z_{hr} = z_h$ by equation:

$$
p' \cdot (\vartheta + \varphi_{hr}) - \sqrt{\rho_{ax}^2 - (K-\eta)^2} + z_{ax} = -p_h \cdot (\vartheta + \varphi_{0h})
\tag{3.144}
$$

In this equation:

$$
\varphi_{hr} = \frac{1}{p'} \cdot \sqrt{\rho_{ax}^2 - (K - \eta^2} - \frac{p_h}{p'} \cdot (\vartheta + \varphi_{0h}) - \vartheta - \frac{z_{ax}}{p'}
\tag{3.145}
$$

As this line of intersection should coincide with the cutting edge determined by (3.142) it is necessary that φ_{on} as given in (3.140) fulfils equation (3.144). That is:

$$
\varphi_{hr} = \frac{1}{p'} \cdot \sqrt{\rho_{ax}^2 - (K-\eta')^2} - \frac{z_{ax}}{p'} + \frac{p_h}{p'} \cdot \left[\frac{p}{p_h}(\vartheta + \varphi_0) - \frac{1}{p_h} \cdot \sqrt{\rho_{ax}^2 - (K-\eta')^2} + \frac{z_{ax}}{p_h} \right] - \vartheta
$$

After simplifying the equation:

$$\varphi_{hr} = \frac{p}{p'}(\vartheta + \varphi_0) - \vartheta. \tag{3.146}$$

The equation of the relieved surface after substituting (3.146) into (3.143) is obtained as:

$$\left. \begin{array}{l} x_{hr} = -\eta \cdot \sin(\frac{p}{p'} \cdot \Theta) \\[2mm] y_{hr} = \eta' \cos(\frac{p}{p'} \cdot \Theta) \\[2mm] z_{hr} = p' \cdot (\frac{p}{p'} \cdot \Theta) - \sqrt{\rho_{ax}^2 - (K - \eta)^2} + z_{ax} \end{array} \right\} \tag{3.147}$$

When $\varphi_1 = 0$ and $x = 0$, the axial section of the back surface is obtained, and this can be used for simple checking calculations for controlling the milling cutter:

$$\left. \begin{array}{l} x_1 = 0 \text{ in case of} : \frac{p}{p'} \cdot \Theta = 0 \\[2mm] x_{hr} = 0; \\[1mm] y_{hr} = \eta'; \\[1mm] z_{hr} = -\sqrt{\rho_{ax}^2 - (K - \eta')^2} + z_{ax} \end{array} \right\} \tag{3.148}$$

From equation (3.146) it can be seen that when $p' = p$, $r_{hr} = r_1$ (worm). Applying the wheel-generating method worked out in this book, switching off the relief and applying the kinematic relation determined by the p' parameter, the vertex of the diamond generates the 'relief' side surface of the worm generator cutter.

The surface of the grinding wheel shaped up this way is a surface of rotation and it is the conjugate surface of the relieved surface.

Determination of the new thread-face surface on fly cutter tool

The range of application of fly cutters is significantly increased when instead of using lands, the total face width of the blade side surface is relieved, following precise geometry. By resharpening the relieved blade along the thread face surface (closed, flat) it can be guaranteed that this tool will generate a correct profile until it is completely used up. The limit for resharpening is determined by strength considerations and the clearances needed at the addendum and at the sides.

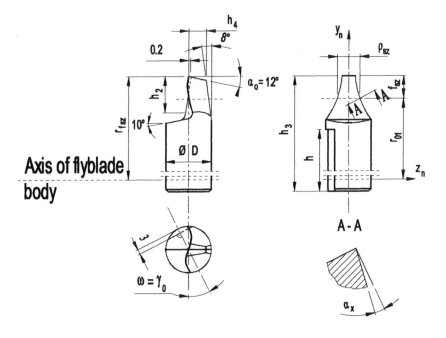

Figure 3.43 *Shape of fly cutter with thread face surface*

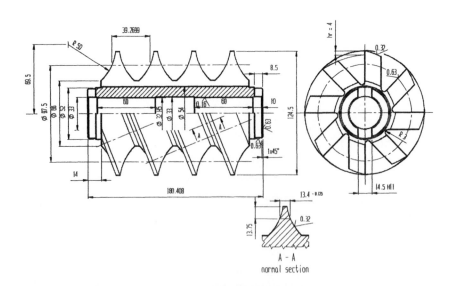

Figure 3.44 *Generating worm milling cutter worm gears with curved profile*

Figure 3.45 *Grinding of worm gear generating tool (fly cutter)*

Figure 3.46 *Relieving a fly cutter or a milling cutter*

The dimensions of the fly cutter profile should be determined as with the milling cutter.

The relief of the tool, for both fly and milling cutters, can be carried out using an end or disc type grinding wheel. In the case of a disc type tool, the diameter of the wheel can be determined only by knowing the geometric shape and the value of the relief (Maros, Killmann and Rohonyi, 1970).

The face angle of the designed tool is $\gamma=0$ so no deformed profile can occur owing to the geometric shape of the milling cutter (Bakondi, 1974).

Investigating the manufactured tool profiles in axial section, it can be stated that from the point of view of precision of shape they are satisfactory. The achieved values of shape precision are less than the allowable values $f_{fr} = 0.03$ mm < 0.05 mm (Dudás, 1998e). The method of assessment is discussed in Chapter 6.

Naturally if applying radial relief after resharpening, the radial sizes will have changed (owing to cutting edge displacement). This effect was investigated in the range $\theta = 0°{-}20°$ (Dudás, 1998e). The sizes of tool calculated in axial section showed very good correlation with values measured by profile measurements. To guarantee the reserve module for resharpening (usual range 0.04–01) no significant profile deviation occurred owing to changes in radial size; the value of it remained less than the allowable profile deviation (Figure 4.42).

The clamping of the manufactured flyblade is illustrated in Figure 3.45. In this construction, the face plane of the tool is perpendicular to the machined surface; because of the equal angles of inclination of the side surfaces, an equilibrium of forces comes into effect which itself is beneficial from the point of view of machining. The geometry of this new fly cutter is shown in Figure 3.43, its relief in Figure 3.46 and its face plane sharpening in Figure 3.47. (A Klingelnberg GW A type machine was used.)

Serious technological difficulties arise in producing a theoretically precise worm-generator milling cutter. This is why approximate solutions are generally chosen to keep errors within tolerance range.

The tools and worm–worm gear pairs manufactured for the purpose of investigation were:

$$a = 125 \text{ mm}; \qquad m = 7.5 \text{ mm} \qquad z_1 = 1 \text{ and}$$
$$a = 280 \text{ mm}; \qquad m = 12.5 \text{ mm}; \qquad z_1 = 3 \text{ and}$$
$$a = 280 \text{ mm} \qquad m = 16 \text{ mm}; \qquad z_1 = 5$$

Figure 3.47 *Resharpening at face plane of a relieved fly cutter*

The generator milling cutter can be created by multiplying the thread-face profile fly cutter as a tooth of the hob. The geometric shape of this hob can be seen in Figure 3.44. Its basic data are i = 11.67, m = 12.5 mm.

4

GENERAL MATHEMATICAL MODEL FOR INVESTIGATION OF HOBS SUITABLE FOR GENERATING CYLINDRICAL AND CONICAL WORMS, WORM GEARS AND FACE GEAR GENERATORS

By bringing together the model created to investigate cylindrical helicoidal surfaces and their tools and that prepared for investigation of conical helicoidal surfaces, a general kinematic model can be obtained (see Figure 4.1), suitable for treatment by a single mathematical model.

The applied coordinate systems and symbols used are equivalent to the ones previously mentioned (p_r the radial thread parameter, c the value of tool displacement).

The transformation matrices are: $\underline{M}_{1,1F}$; $\underline{M}_{1F,1}$ as (3.99), $\underline{M}_{0,1}$; $\underline{M}_{1,0}$ as (3.100) and $\underline{M}_{2F,2}$; $\underline{M}_{2,2F}$ as determined in (3.102).

According to Figure 4.1:

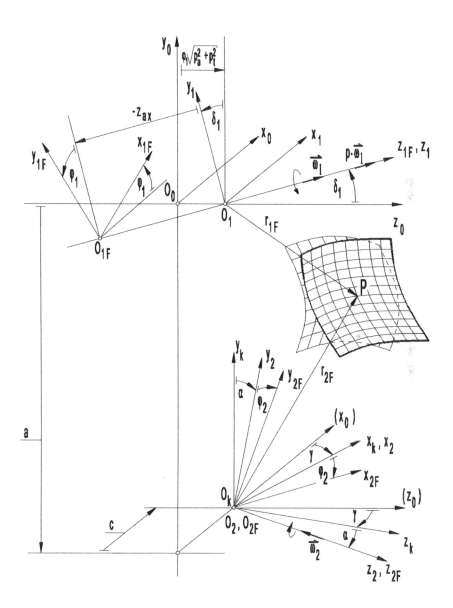

Figure 4.1 _Correlation between coordinate systems for general investigation of machining theory of cylindrical and conical helicoidal surfaces_

$$
\underline{M}_{K,0} = \begin{bmatrix} \cos\gamma & 0 & \sin\gamma & -c\cdot\cos\gamma \\ 0 & 1 & 0 & a \\ -\sin\gamma & 0 & \cos\gamma & c\cdot\sin\gamma \\ 0 & 0 & 0 & 1 \end{bmatrix}; \quad \underline{M}_{0,K} = \begin{bmatrix} \cos\gamma & 0 & -\sin\gamma & c \\ 0 & 1 & 0 & -a \\ \sin\gamma & 0 & \cos\gamma & 0 \\ 0 & 0 & 0 & 1 \end{bmatrix}; \quad (4.1)
$$

$$
\underline{M}_{2,K} = \begin{bmatrix} 1 & 0 & 0 & 0 \\ 0 & \cos\gamma & \sin\gamma & 0 \\ 0 & -\sin\gamma & \cos\gamma & 0 \\ 0 & 0 & 0 & 1 \end{bmatrix}; \quad \underline{M}_{K,2} = \begin{bmatrix} 1 & 0 & 0 & 0 \\ 0 & \cos\gamma & -\sin\gamma & 0 \\ 0 & \sin\gamma & \cos\gamma & 0 \\ 0 & 0 & 0 & 1 \end{bmatrix}; \quad (4.2)
$$

The direct transformation matrices between rotating coordinate systems are:

$$
\underline{M}_{2F,1F} = \underline{M}_{2F,2}\cdot\underline{M}_{2,K}\cdot\underline{M}_{K,0}\cdot\underline{M}_{0,1}\cdot\underline{M}_{1,1F}, \text{ or} \quad (4.3)
$$

$$
\underline{M}_{1F,2F} = \underline{M}_{1F,1}\cdot\underline{M}_{1,0}\cdot\underline{M}_{0,K}\cdot\underline{M}_{K,2}\cdot\underline{M}_{2,2F} \quad (4.4)
$$

The solution of the direct task (surface of workpiece is known) when knowing $\vec{r}_{1F}$, the surface No2 and point of contact line is sought using equations (3.72), (3.74), (3.75), (3.76), (3.77), (3.78) and (3.79). Further, the matrices depending only on a kinematic relation can be determined as:

$$
\frac{d\underline{M}_{2F,1F}}{dt} \quad (4.5)
$$

and:

$$
P_{1a} = \underline{M}_{1F,2F}\cdot\frac{d\underline{M}_{2F,1F}}{dt} \quad (4.6)
$$

After completing the operations, the matrix (4.13) is obtained. P_{1a} is the matrix for the kinematic generation of the general model. In solving the inverse problem, only the direction of transformation changes into the opposite one.

Known:

$$
\vec{r}_{2F} = \vec{r}_{2F}(y_{20}, \Psi) \quad (4.7)
$$

The surface No1 according to the theory of enveloping surfaces can be derived by differentiating with respect to the movement parameter the series of surfaces generated by $\vec{r}_{2F}$ during its movement, while contact curve can be obtained by the simultaneous solution of equations:

$$\vec{n}_{2F} \cdot \vec{v}_{2F}^{21} = 0 \quad \text{and} \quad \vec{r}_{1F} = \underline{M}_{1F,2F} \cdot \vec{r}_{2F} \tag{4.8}$$

in the K_{1F} coordinate system, where:

$$\vec{n}_{2F} = \frac{\partial \vec{r}_{2F}}{\partial y_{20}} \times \frac{\partial \vec{r}_{2F}}{\partial \Psi} \qquad \vec{v}_{2F}^{(21)} = \underline{P}_{2a} \cdot \vec{r}_{2F} \tag{4.9}$$

To obtain the solution of this problem the P_{2a} matrix should be determined as:

$$P_{2a} = \underline{M}_{2F,1F} \cdot \frac{d\underline{M}_{1F,2F}}{dt} \tag{4.10}$$

The matrices are defined in the $\underline{M}_{1F,2F}$ equation (4.11), and $\underline{M}_{2F,1F}$ in (4.12).

The derived matrix $\left(\dfrac{d}{dt} \underline{M}_{1F,2F} \right)$ is derived to use for generating the general kinematic conditions for the coordinate systems given in Figure 4.1 similarly to (3.80–3.85). The kinematic generation matrices $\underline{P}_{1a}$ and $\underline{P}_{2a}$ are given by equations (4.13) and (4.14).

The model may be used to investigate conditions of meshing for both conical and cylindrical helicoidal surfaces with the tool having its body as the block of rotation.

Using this model it is possible to determine the contact curve starting from a known $\vec{r}_{1F}$ (No1 workpiece) (the so-called direct problem solution) as well as starting from known $\vec{r}_{2F}$ (No2 tool). We shall see further on in the text that it is also possible to start from a given contact curve as a generating curve to determine the tool surface No2 according to equation (3.79) and the surface of workpiece No1 as equation (3.85) describes it.

The surface of workpiece No1 is a cylindrical or conical basic surface having a thread curve fitted with an arbitrary generator curve (the axial section of thread).

For surface of tool No2 it is useful to define a surface of rotation, but it can be any other surface too, for example a single cutting edge tool with a cutting edge determined by $\varphi_2 = $ const. The frequently used types of workpiece and tool are shown in Figure 4.2 in a tabular form giving the value of kinematic parameters which in some cases can be equal to 0.

In section 3.3, while analysing the machining geometry of generator milling cutters, several types of possible back surfaces were determined, and it was concluded that the axial back surface can be regarded as cylindrical while radial and diagonal types of back surfaces can be handled as conical helicoid surfaces.

$$\underline{M}_{1F,2F} = \underline{M}_{1F,0} \cdot \underline{M}_{0,2F} =$$

$$
\begin{bmatrix}
\begin{aligned}
&\cos\gamma\cdot\cos\varphi_1\cdot\cos\varphi_2 + \\
&+\sin\alpha\cdot\sin\gamma\cdot\cos\varphi_1\cdot\sin\varphi_2 - \\
&-\cos\alpha\cdot\cos\delta\cdot\sin\varphi_1\cdot\sin\varphi_2 - \\
&-\sin\gamma\cdot\sin\delta\cdot\sin\varphi_1\cdot\cos\varphi_2 + \\
&+\sin\alpha\cdot\cos\gamma\cdot\sin\delta\cdot\sin\varphi_1\cdot\sin\varphi_2
\end{aligned}
&
\begin{aligned}
&\cos\gamma\cdot\cos\varphi_1\cdot\sin\varphi_2 - \\
&-\sin\alpha\cdot\sin\gamma\cdot\cos\varphi_1\cdot\cos\varphi_2 + \\
&+\cos\alpha\cdot\cos\delta\cdot\sin\varphi_1\cdot\cos\varphi_2 - \\
&-\sin\gamma\cdot\sin\delta\cdot\sin\varphi_1\cdot\sin\varphi_2 - \\
&-\sin\alpha\cdot\cos\gamma\cdot\sin\delta\cdot\sin\varphi_1\cdot\cos\varphi_2
\end{aligned}
&
\begin{aligned}
&-\cos\alpha\cdot\sin\gamma\cdot\cos\varphi_1 - \\
&-\sin\alpha\cdot\cos\delta\cdot\sin\varphi_1 - \\
&-\cos\alpha\cdot\cos\gamma\cdot\sin\delta\cdot\sin\varphi_1
\end{aligned}
&
\begin{aligned}
&c\cdot\cos\varphi_1 - \\
&-a\cdot\cos\delta\cdot\sin\varphi_1 + \\
&+p_r\cdot\varphi_1\cdot\sin\varphi_1
\end{aligned}
\\[2em]
\begin{aligned}
&-\cos\gamma\cdot\sin\varphi_1\cdot\cos\varphi_2 - \\
&-\sin\alpha\cdot\sin\gamma\cdot\sin\varphi_1\cdot\sin\varphi_2 - \\
&-\cos\alpha\cdot\cos\delta\cdot\cos\varphi_1\cdot\sin\varphi_2 - \\
&-\sin\gamma\cdot\sin\delta\cdot\cos\varphi_1\cdot\cos\varphi_2 + \\
&+\sin\alpha\cdot\cos\gamma\cdot\sin\delta\cdot\cos\varphi_1\cdot\sin\varphi_2
\end{aligned}
&
\begin{aligned}
&-\cos\gamma\cdot\sin\varphi_1\cdot\sin\varphi_2 + \\
&+\sin\alpha\cdot\sin\gamma\cdot\sin\varphi_1\cdot\cos\varphi_2 + \\
&+\cos\alpha\cdot\cos\delta\cdot\cos\varphi_1\cdot\cos\varphi_2 - \\
&-\sin\gamma\cdot\sin\delta\cdot\cos\varphi_1\cdot\sin\varphi_2 - \\
&-\sin\alpha\cdot\cos\gamma\cdot\sin\delta\cdot\cos\varphi_1\cdot\cos\varphi_2
\end{aligned}
&
\begin{aligned}
&\cos\alpha\cdot\sin\gamma\cdot\sin\varphi_1 - \\
&-\sin\alpha\cdot\cos\delta\cdot\cos\varphi_1 - \\
&-\cos\alpha\cdot\cos\gamma\cdot\sin\delta\cdot\cos\varphi_1
\end{aligned}
&
\begin{aligned}
&-c\cdot\sin\varphi_1 - \\
&-a\cdot\cos\delta\cdot\cos\varphi_1 + \\
&+p_r\cdot\varphi_1\cdot\cos\varphi_1
\end{aligned}
\\[2em]
\begin{aligned}
&-\cos\alpha\cdot\sin\delta\cdot\sin\varphi_2 + \\
&+\sin\gamma\cdot\cos\delta\cdot\cos\varphi_2 - \\
&-\sin\alpha\cdot\cos\gamma\cdot\cos\delta\cdot\sin\varphi_2
\end{aligned}
&
\begin{aligned}
&\cos\alpha\cdot\sin\delta\cdot\cos\varphi_2 + \\
&+\sin\gamma\cdot\cos\delta\cdot\sin\varphi_2 + \\
&+\sin\alpha\cdot\cos\gamma\cdot\cos\delta\cdot\cos\varphi_2
\end{aligned}
&
\begin{aligned}
&-\sin\alpha\cdot\sin\delta + \\
&+\cos\alpha\cdot\cos\gamma\cdot\cos\delta
\end{aligned}
&
\begin{aligned}
&-a\cdot\sin\delta - \\
&-p_r\cdot\varphi_1 - z_{ax}
\end{aligned}
\\[2em]
0 & 0 & 0 & 1
\end{bmatrix}
$$

$$(4.11)$$

$$\underline{M}_{2F.1F} = \underline{M}_{2F.K} \cdot \underline{M}_{K.1F} =$$

$$
\begin{bmatrix}
\begin{aligned}&\cos\gamma\cos\varphi_1\cos\varphi_2 - \\ &-\sin\gamma\sin\delta\sin\varphi_1\cos\varphi_2 - \\ &-\cos\alpha\cos\delta\sin\varphi_1\sin\varphi_2 + \\ &+\sin\alpha\sin\gamma\cos\varphi_1\sin\varphi_2 + \\ &+\sin\alpha\cos\gamma\sin\delta\sin\varphi_1\cdot \\ &\cdot\sin\varphi_2\end{aligned}
&
\begin{aligned}&-\cos\gamma\sin\varphi_1\cos\varphi_2 - \\ &-\sin\gamma\sin\delta\sin\varphi_1\cos\varphi_2 - \\ &-\cos\alpha\cos\delta\cos\varphi_1\sin\varphi_2 - \\ &-\sin\alpha\sin\gamma\sin\varphi_1\sin\varphi_2 + \\ &+\sin\alpha\cos\gamma\sin\delta\sin\varphi_1\cdot \\ &\sin\varphi_2\end{aligned}
&
\begin{aligned}&\sin\gamma\cos\delta\cos\varphi_2 - \\ &-\cos\alpha\sin\delta\sin\varphi_2 - \\ &-\sin\alpha\cos\gamma\cos\delta\sin\varphi_2\end{aligned}
&
\begin{aligned}&(z_{ax}\sin\gamma\cos\delta - c\cdot\cos\gamma)\cos\varphi_2 + \\ &+\sqrt{p_a^2 + p_r^2}\,\sin\gamma\cdot\varphi_1\cos\varphi_2 - \\ &-(z_{ax}\sin\delta + a)\cos\alpha\sin\varphi_2 - \\ &-(z_{ax}\cos\gamma\cos\delta + c\cdot\sin\gamma)\sin\alpha\sin\varphi_2 - \\ &-\sqrt{p_a^2 + p_r^2}\,\sin\alpha\cos\gamma\cdot\varphi_1\sin\varphi_2\end{aligned}
\\[2em]
\begin{aligned}&\cos\gamma\cos\varphi_1\sin\varphi_2 - \\ &-\sin\gamma\sin\delta\sin\varphi_1\sin\varphi_2 + \\ &+\cos\alpha\cos\delta\sin\varphi_1\cos\varphi_2 - \\ &-\sin\alpha\sin\gamma\cos\varphi_1\cos\varphi_2 - \\ &-\sin\alpha\cos\gamma\sin\delta\sin\varphi_1\cdot \\ &\cos\varphi_2\end{aligned}
&
\begin{aligned}&-\cos\gamma\sin\varphi_1\sin\varphi_2 - \\ &-\sin\gamma\sin\delta\cos\varphi_1\sin\varphi_2 + \\ &+\cos\alpha\cos\delta\cos\varphi_1\cos\varphi_2 - \\ &-\sin\alpha\sin\gamma\cos\varphi_1\cos\varphi_2 - \\ &-\sin\alpha\cos\gamma\sin\delta\sin\varphi_1\cdot \\ &\cos\varphi_2\end{aligned}
&
\begin{aligned}&\sin\gamma\cos\delta\sin\varphi_2 + \\ &+\cos\alpha\sin\delta\cos\varphi_2 + \\ &+\sin\alpha\cos\gamma\cos\delta\cos\varphi_2\end{aligned}
&
\begin{aligned}&(z_{ax}\sin\gamma\cos\delta - c\cdot\cos\gamma)\sin\varphi_2 + \\ &+\sqrt{p_a^2 + p_r^2}\,\sin\gamma\cdot\varphi_1\sin\varphi_2 + \\ &+(z_{ax}\sin\delta + a)\cos\alpha\cos\varphi_2 - \\ &-(z_{ax}\cos\gamma\cos\delta + c\cdot\sin\gamma)\sin\alpha\cos\varphi_2 - \\ &-\sqrt{p_a^2 + p_r^2}\,\sin\alpha\cos\gamma\cdot\varphi_1\cos\varphi_2\end{aligned}
\\[2em]
\begin{aligned}&-\sin\alpha\cos\delta\sin\varphi_1 - \\ &-\cos\alpha\sin\gamma\cos\varphi_1 - \\ &-\cos\alpha\cos\gamma\sin\delta\sin\varphi_1\end{aligned}
&
\begin{aligned}&-\sin\alpha\cos\delta\cos\varphi_1 + \\ &+\cos\alpha\sin\gamma\cos\varphi_1 - \\ &-\cos\alpha\cos\gamma\sin\delta\sin\varphi_1\end{aligned}
&
\begin{aligned}&-\sin\alpha\sin\delta + \\ &+\cos\alpha\cos\gamma\cos\delta\end{aligned}
&
\begin{aligned}&(z_{ax}\sin\delta + a)\sin\alpha + \\ &+(z_{ax}\cos\gamma\cos\delta + c\cdot\sin\gamma)\cos\alpha + \\ &+\sqrt{p_a^2 + p_r^2}\,\cos\alpha\cos\gamma\cdot\varphi_1\end{aligned}
\\[2em]
0 & 0 & 0 & 1
\end{bmatrix}
$$

$$(4.12)$$

$$P_{1a} =
\begin{bmatrix}
\begin{aligned}&\big(-c.i.\cos\alpha\cos\gamma\cos\delta - \sqrt{P_a^2+P_r^2}.\sin\delta + i.z_{ax}\cos\alpha\sin\gamma + c.i.\sin\alpha\sin\delta + a.i.\cos\alpha\sin\gamma\sin\delta\big)\sin\varphi_1 \\ &- \big(i.z_{ax}.\sin\alpha\cos\delta + i.z_{ax}\cos\alpha\cos\gamma\sin\delta + a.i.\cos\alpha\cos\gamma\big)\cos\varphi_1 \\ &+ i\sqrt{P_a^2+P_r^2}.\cos\alpha\sin\gamma\cos\delta.\varphi_1\sin\varphi_1 - i\sqrt{P_a^2+P_r^2}.\sin\alpha.\varphi_1\cos\varphi_1\end{aligned}
&
\begin{aligned}&-i.\sin\alpha\cos\delta\cos\varphi_1 \\ &-i.\cos\alpha\cos\gamma\sin\delta\cos\varphi_1 \\ &+i.\cos\alpha\sin\gamma\sin\varphi_1\end{aligned}
&
\begin{aligned}&-1-i.\cos\alpha\cos\gamma\cos\delta \\ &+i.\sin\alpha\sin\delta;\end{aligned}
& 0 \\[2em]

\begin{aligned}&\big(i.z_{ax}\sin\alpha\cos\delta + i.z_{ax}\cos\alpha\cos\gamma\sin\delta + a.i.\cos\alpha\cos\gamma\big).\sin\varphi_1 \\ &+ i.z_{ax}\cos\alpha\sin\gamma + a.i.\cos\alpha\sin\gamma\sin\delta + c.i.\sin\alpha\sin\delta \\ &- c.i.\cos\alpha\cos\gamma\cos\delta - \sqrt{P_a^2+P_r^2}.\sin\delta\cos\varphi_1 \\ &+ i\sqrt{P_a^2+P_r^2}.\sin\alpha.\varphi_1\sin\varphi_1 \\ &+ i\sqrt{P_a^2+P_r^2}.\cos\alpha\sin\gamma\cos\delta.\varphi_1\cos\varphi_1\end{aligned}
&
\begin{aligned}&i.\sin\alpha\cos\delta\sin\varphi_1 \\ &+i.\cos\alpha\cos\gamma\sin\delta\sin\varphi_1 \\ &+i.\cos\alpha\sin\gamma\cos\varphi_1\end{aligned}
&
\begin{aligned}&1+i.\cos\alpha\cos\gamma\cos\delta \\ &-i.\sin\alpha\sin\delta\end{aligned}
& 0 \\[2em]

\begin{aligned}&i.\sqrt{P_a^2+P_r^2}.\cos\alpha\sin\gamma\sin\delta\,\varphi_1 \\ &- c.i.\cos\alpha\cos\gamma\sin\delta + \sqrt{P_a^2+P_r^2}.\cos\delta \\ &- a.i.\cos\alpha\sin\gamma\cos\delta - c.i.\sin\alpha\cos\delta\end{aligned}
& 0 & 0 & 0 \\[2em]

\begin{aligned}&i.\sin\alpha\cos\delta\cos\varphi_1 \\ &+i.\cos\alpha\cos\gamma\sin\delta\cos\varphi_1 \\ &-i.\cos\alpha\sin\gamma\sin\varphi_1\end{aligned}
&
\begin{aligned}&-i.\sin\alpha\cos\delta\sin\varphi_1 \\ &-i.\cos\alpha\cos\gamma\sin\delta\sin\varphi_1 \\ &-i.\cos\alpha\sin\gamma\cos\varphi_1\end{aligned}
& 0 & 0
\end{bmatrix}$$

$$(4.13)$$

$$
\underline{P}_{2a} =
\begin{bmatrix}
0 & i - \sin\alpha\sin\delta + \cos\alpha\cos\gamma\cos\delta & -\sin\gamma\cos\delta\sin\varphi_2 - (\cos\alpha\sin\delta + \sin\alpha\cos\gamma\cos\delta)\cos\varphi_2 & \begin{aligned}&(-p_r\cos\alpha\cos\delta + c.\cos\alpha\cos\delta + p_r\sin\alpha\cos\gamma\sin\delta - \\ &c.\sin\alpha\cos\gamma\sin\delta - a.\sin\alpha\sin\gamma\cos\delta + \\ &p_a\cos\alpha\sin\delta + p_a\sin\alpha\cos\gamma\cos\delta)\sin\varphi_2 + \\ &(c.\sin\gamma\sin\delta - p_r\sin\gamma\sin\delta - a.\cos\gamma\cos\delta - \\ &p_a\sin\gamma\cos\delta)\cos\varphi_2 + \\ &p_r\sin\alpha\sin\gamma.\varphi_1\sin\varphi_2 + p_r\cos\gamma.\varphi_1\cos\varphi_2\end{aligned} \\
-i + \sin\alpha\sin\delta - \cos\alpha\cos\gamma\cos\delta & 0 & (-\cos\alpha\sin\delta - \sin\alpha\cos\gamma\cos\delta)\sin\varphi_2 + \sin\gamma\cos\delta\cos\varphi_2 & \begin{aligned}&(-p_r\sin\gamma\sin\delta + c.\sin\gamma\sin\delta - a.\cos\gamma\cos\delta - \\ &p_a\sin\gamma\cos\delta)\sin\varphi_2 + (p_r\cos\alpha\cos\delta - c.\cos\alpha\cos\delta - p_r\sin\alpha\cos\gamma\sin\delta + \\ &c.\sin\alpha\cos\gamma\sin\delta + a.\sin\alpha\sin\gamma\cos\delta - \\ &p_a\cos\alpha\sin\delta - p_a\sin\alpha\cos\gamma\cos\delta)\cos\varphi_2 + \\ &p_r\cos\gamma.\varphi_1\sin\varphi_2 - p_r\sin\alpha\sin\gamma.\varphi_1\cos\varphi_2\end{aligned} \\
\begin{aligned}&\sin\gamma\cos\delta\sin\varphi_2 + \\ &(\sin\alpha\cos\gamma\cos\delta + \\ &\cos\alpha\sin\delta)\cos\varphi_2\end{aligned} & (\sin\alpha\cos\gamma\cos\delta + \cos\alpha\sin\delta)\sin\varphi_2 - \sin\gamma\cos\delta\cos\varphi_2 & 0 & \begin{aligned}&-p_r\cos\alpha\sin\gamma.\varphi_1 + \\ &c.\sin\alpha\cos\delta - p_r\sin\alpha\cos\delta - p_r\cos\alpha\cos\gamma\sin\delta + \\ &c.\cos\alpha\cos\gamma\sin\delta + a.\cos\alpha\sin\gamma\cos\delta + \\ &p_a\sin\alpha\sin\delta - p_a\cos\alpha\cos\gamma\cos\delta\end{aligned} \\
0 & 0 & 0 & 0
\end{bmatrix}
$$

$$(4.14)$$

Movement geometrical feature									
Tool type is 2		Type of workpiece 1.	a	c	α	γ	δ	P_a	P_r
Disc type milling	Cylindrical worm	ZA	>0	0	0	≠0	0	≠0	0
		ZI[*]	>0	0	0	≠0	0	≠0	0
		ZI[**]	>0	≠0	>0	≠0	0	≠0	0
		ZN	>0	0	0	≠0	0	≠0	0
		ZT	>0	0	0	≠0	0	≠0	0
		ZK	0	0	0	≠0	0	≠0	0
	Conical worm	KA	>0	0	0	≠0	>0	≠0	>0
		KI[*]	>0	0	0	≠0	>0	≠0	>0
		KI[**]	>0	≠0	>0	≠0	>0	≠0	>0
		KN	>0	0	0	≠0	>0	≠0	>0
		KT	>0	0	0	≠0	>0	≠0	>0
		KK	>0	0	0	≠0	>0	≠0	>0
		Axial flank surface	>0	0	0	≠0	0	≠0	0
		Radial and diagonal flank surface	>0	0	0	≠0	>0	≠0	>0
Pin type grinding	Cylindrical worm	ZA	>0	0	-90°	0	0	≠0	0
		ZI[*]	>0	0	-90°	0	0	≠0	0
		ZI[**]	-	-	-	-	-	-	-
		ZN	>0	0	-90°	0	0	≠0	0
		ZT	>0	0	-90°	0	0	≠0	0
		ZK	>0	0	-90°	0	0	≠0	0
	Conical worm	KA	>0	0	-90°	0	>0	≠0	>0
		KI[*]	>0	0	-90°	0	>0	≠0	>0
		KI[**]	-	-	-	-	-	-	-
		KN	>0	0	-90°	0	>0	≠0	>0
		KT	>0	0	-90°	0	>0	≠0	>0
		KK	>0	0	-90°	0	0	≠0	0
		Axial flank surface	0	0	-90°	0	0	≠0	0
		Radial and diagonal flank surface	0	0	-90°	0	>0	≠0	>0
Straight cup	Cylindrical worm	ZA	>0	≠0	>0	≠0	>0	≠0	0
		ZI[*]	>0	≠0	>0	≠0	0	≠0	0
		ZI[**]	-	-	-	-	-	-	-
		ZN	>0	≠0	>0	≠0	0	≠0	0
		ZT	>0	≠0	>0	≠0	0	≠0	0
		ZK	>0	≠0	>0	≠0	0	≠0	0
	Conical worm	KA	>0	≠0	>0	≠0	>0	≠0	>0
		KI[*]	>0	≠0	>0	≠0	>0	≠0	>0
		KI[**]	-	-	-	-	-	-	-
		KN	>0	≠0	>0	≠0	>0	≠0	>0
		KT	>0	≠0	>0	≠0	>0	≠0	>0
		KK	>0	≠0	>0	≠0	>0	≠0	>0
		Axial flank surface	>0	≠0	>0	≠0	0	≠0	0
		Radial and diagonal flank surface	>0	≠0	>0	≠0	>0	≠0	>0

* With tapered grinding wheel
** Manufacturing with straight grinding wheel in a lifted out position

Figure 4.2 *The most frequently used types of workpiece and tool surfaces characterized by general model parameters*

Later in this chapter the determination of the edge curve as the generator curve of the back surface (in the case of a cylindrical worm, having an axial section circular profile), is discussed and there is also shown the generation of radial back surface determining the equation of the contact curve between the radial back surface and the finishing grinding tool.

For the purpose of comparison with the conical model during machining, the disc-like wheel was used.

Figure 4.3c presents a typical example of the machining positions shown in Figure 4.2. The determination of the parameters in these two cases (Figure 4.3a, b.) and the presentation in the figure are found by applying the general model. A great number of variants could be envisaged, but within the framework of this book it is impossible to review them all, and the following is one example that can be calculated.

For many of the technological and kinematic positions illustrated in Figures 4.2 and 4.3 it can be shown that the transformation $(\underline{M}_{1F,2F}, \underline{M}_{2F,1F})$ and the kinematic generation matrices $(\underline{P}_{1a}, \underline{P}_{2a})$ created for the general model contain all possible situations.

So, by substitution of suitable parameters, matrices can be obtained both for cylindrical and for conical worm gear drives (see Chapter 3). Naturally, matrices valid for other positions can also be generated as well.

For example, let us consider the generation of a few matrix elements using the general model.

■ Grinding of cylindrical worm:

$$\delta = 0, \; p_r = 0, \; c = 0, \; \alpha = 0$$

Substituting them into the appropriate element of (4.11) the element No2 in the first row of the matrix can be obtained as:

$$\underline{M}_{2F,1F}^{(1,2)} = -\cos\gamma\sin\varphi_1\cos\varphi_2 - \sin\varphi_2\cos\varphi_1$$

$$\sin\delta = 0, \qquad \cos\delta = 1$$
$$\sin\alpha = 0, \qquad \cos\alpha = 1$$

This is the element equivalent to the appropriate element of matrix (3.33).

■ Grinding of conical worm:

$$\delta > 0, \; p_r > 0, \; c = 0, \; \alpha = 0$$

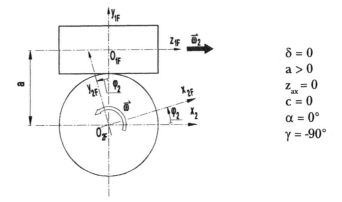

a) Model of cylindrical worm gearing

$\delta = 0$
$a > 0$
$z_{ax} = 0$
$c = 0$
$\alpha = 0°$
$\gamma = -90°$

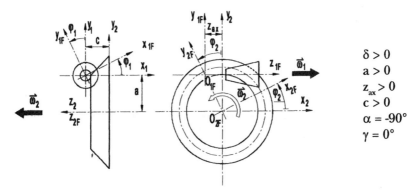

b) Model of spiroid worm gearing

$\delta > 0$
$a > 0$
$z_{ax} > 0$
$c > 0$
$\alpha = -90°$
$\gamma = 0°$

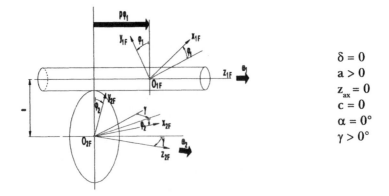

c) Grinding model of cylindrical worm

$\delta = 0$
$a > 0$
$z_{ax} = 0$
$c = 0$
$\alpha = 0°$
$\gamma > 0°$

Figure 4.3 *Main areas of application of general model*

Substituting them into an appropriate element of matrix (4.11) element No2 in the first row of the matrix can be obtained as:

$$\underline{M}^{(1,2)}_{2F,1F} = \left(\sin\delta\sin\gamma\cos\varphi_2 - \cos\delta\sin\varphi_2\right)\cos\varphi_1 - \cos\gamma\cos\varphi_2\sin\varphi_1)$$

$\sin\alpha = 0$, $\cos\alpha = 1$

This is the element equivalent to the appropriate element of matrix (3.112).

■ Investigating in the same case one–one element of matrices $\underline{P}_{1a}$ and $\underline{P}_{1k}$ (for example $\underline{P}^{(2,3)}_{1a}$ it should be:

$$\underline{P}^{(2,3)}_{1a} = i\left(\cos\gamma\sin\delta\sin\varphi_1 + \sin\gamma\cos\varphi\right)$$

This is the element equal to the appropriate element matrix P_{1k} (3.114).

4.1 APPLICATION OF GENERAL MATHEMATICAL – KINEMATIC MODEL TO DETERMINE SURFACE OF HELICOIDAL SURFACE-GENERATING TOOL FOR CYLINDRICAL THREAD SURFACES

In this section results will be presented for cylindrical worm types obtained by applying the general model described above. The tabulation given in Figure 4.4, in its first column, lists the types of worm investigated. The second column contains the parameters substituted into the kinematic sketch presented in Figure 4.1 to obtain the necessary position for generating the given type of worm, shown in column 3.

Because the equations for thread surfaces generated in this way have already been given for tools having line or circular profile, they were repeated. The aim was that, by using the general direct model for all the types given in tabulation of Figure 4.4, to determine the profile of the grinding wheel, depending on the given helicoidal surface for the grinding position ZK1.

Using the formulae of matrices $\underline{P}_{1a}$ (4.12) and $\underline{P}_{2a}$ (4.13) obtained from the general model, and using parameters referring to the method of manufacture and the equations of the known helicoidal surface, the wheel and/or tool profiles given in figures belonging to helicoidal surfaces listed in the tabulation of Figure 4.4 were determined.

The equations obtained by mutual application of the general conical helicoidal surface and the general mathematical–kinematic model obtained after substituting appropriate parameters – the direct geometric sizes – position of generator line, parameters of lead etc, and the points of wheel profile, can be calculated. Similarly, the parameters for the profile function of the wheel for further types of thread listed in the tabulation given in Figure 4.4 as well as for helicoid surfaces having circular profile in axial section can be substituted. When cases differ from these, as is very rare in practice, there is a way of applying the general kinematic matrix P_{1a}, using appropriate parameters, and the equation of helicoidal surface to produce quickly the profile of the wheel (eg tools for machining worm-compressor elements, boring tools with spiral grooves, milling cutters, thread cutters, ball-spindles, guiding-spindles, etc).

In section 4.2 all the above mentioned things are presented as examples of circular profile in an axial section helicoidal surface. In section 4.3 several examples illustrate the determination of a machining wheel profile for ruled conical helicoid surfaces.

Naturally, by applying the inverse method to the helicoidal surface, determination is also possible knowing the tool profile and the relative kinematic position (for example as shown in Figures 4.4a, 4.4e and 4.4f) and their solution is shown in Figs. 4.45 and 4.46.

4.2 MACHINING GEOMETRY OF CYLINDRICAL WORM GEAR DRIVE HAVING CIRCULAR PROFILE IN AXIAL SECTION

The shaping of a helicoidal surface using a grinding wheel can be regarded as the meshing of worm and worm wheel when the angle crossing the angle between their axes is equal to the angle of inclination γ of the wheel (Krivenko, 1967).

Equations or the characteristic describing the grinding wheel as an enveloping surface can be determined as previously discussed. To transfer motion, the meshing side surfaces should remain in continuous contact. Because such surfaces when in relative motion are determined by the same independent parameter, the method to determine the enveloping surface of a single parameter series of surfaces has to be investigated.

At least three different coordinate systems have to be chosen to

Type   Parameters   Connection of coordinate systems; origination of surfaces

ZA

$a = 0$
$b = z_{ax}$
$c = 0$
$\alpha = 0$
$\gamma = 0$
$\varphi = 0$
$\delta = 0$

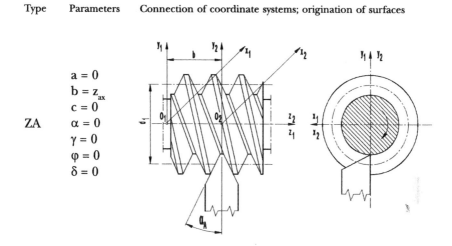

Figure 4.4a

ZN1

$a = 0$
$b = z_{ax}$
$c = 0$
$\alpha = 0$
$\gamma = \gamma_1$
$\varphi_2 = 0$
$\delta = 0$

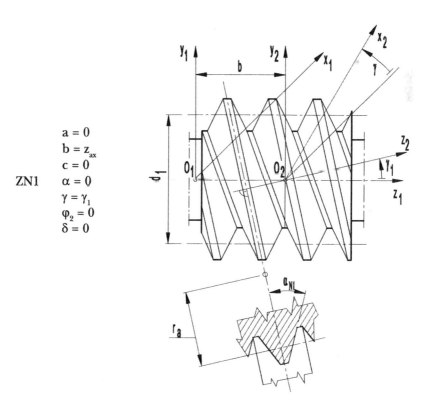

Figure 4.4b

Type   Parameters   Connection of coordinate systems; origination of surface

ZN2

$a = 0$
$b = z_{ax}$
$c = 0$
$\alpha = 0$
$\gamma = \gamma_1$
$\varphi_2 = 0$
$\delta = 0$

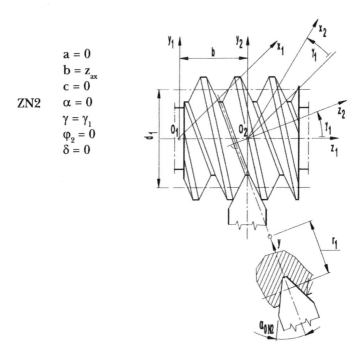

Figure 4.4c

Z1

$a = 0$
$b = z_{ax}$
$c = r_a$
$\alpha = 0$
$\gamma = 0$
$\varphi_2 = 0$
$\delta = 0$

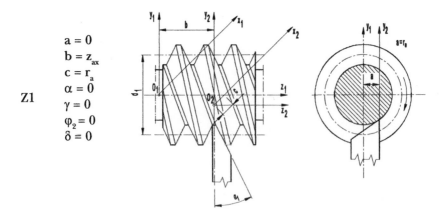

Figure 4.4d

Type   Parameters   Connection of coordinate systems; origination of surface

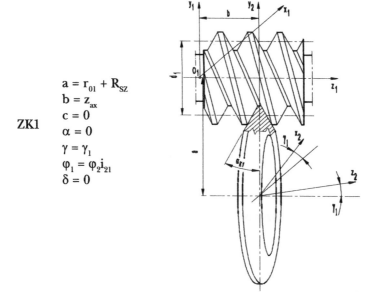

ZK1

$$a = r_{01} + R_{SZ}$$
$$b = z_{ax}$$
$$c = 0$$
$$\alpha = 0$$
$$\gamma = \gamma_1$$
$$\varphi_1 = \varphi_2 i_{21}$$
$$\delta = 0$$

Figure 4.4e

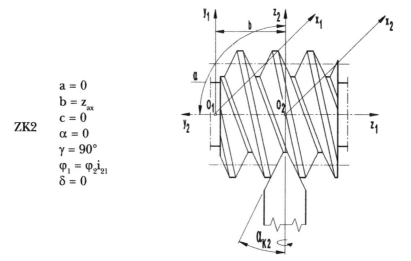

ZK2

$$a = 0$$
$$b = z_{ax}$$
$$c = 0$$
$$\alpha = 0$$
$$\gamma = 90°$$
$$\varphi_1 = \varphi_2 i_{21}$$
$$\delta = 0$$

Figure 4.4f

Type   Parameters   Connection of coordinate systems; origination of surface

ZTA $a = 0$
$b = z_{ax}$
$c = 0$
$\alpha = 0$
$\gamma = \gamma_1$
$\delta = 0$

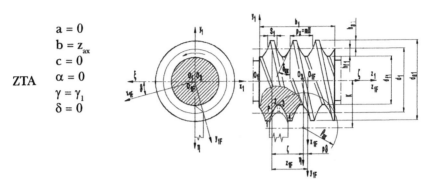

Figure 4.4g

Figure 4.4 *Parameters used for generation of cylindrical worm gear drives in accordance with general mathematical–kinematic model*

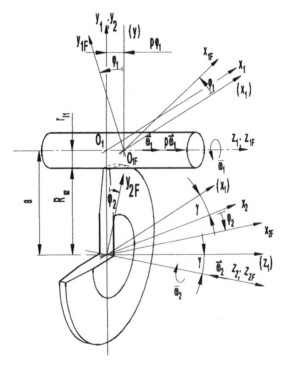

Figure 4.5 *Investigation of grinding; correlation between coordinate systems*

$a = r_{11} + R_K$; $b = 0$; $c = 0$; $\alpha = 0$; $\gamma > 0$; $\varphi_2 = i_{21} \times \varphi_1$; $\delta = 0$

investigate the drive between skew axes (Figure 4.5). In this case four coordinate systems are used for the general model (Figure 4.1).

All the applied coordinate systems are right-hand systems. They are defined as shown in Figure 4.1.

Substituting $\delta = 0$; $\alpha = 0$, the values obtained from Figures 4.5 and 4.1, into the general matrix $\underline{P}_{1h}$ (equation (3.86)), the matrix (3.45) can be obtained. This verifies that it is possible to deduce this specific case from the general mathematical model.

Knowing matrix $\underline{P}_{1h}$ (which is equal to matrix (3.45)) the vector of relative velocity using equation (3.39) can be calculated:

$$
\left.\begin{aligned}
v_{1Fx}^{(12)} &= -\eta\cos\vartheta(1+i_{21}\cos\gamma)+i_{21}\sin\gamma\sin\varphi_1\left[p\vartheta-\sqrt{\rho_{ax}^{2}-(K-\eta)^2}+c+p\varphi_1\right]-i_{21}a\cos\gamma\cos\varphi_1 \\
v_{1Fy}^{(12)} &= -\eta\sin\vartheta(1+i_{21}\cos\gamma)+i_{21}\sin\gamma\cos\varphi_1\left[p\vartheta-\sqrt{\rho_{ax}^{2}-(K-\eta)^2}+c+p\varphi_1\right]+i_{21}a\cos\gamma\sin\varphi_1 \\
v_{1Fz}^{(12)} &= -\eta i_{21}\sin\gamma\cos(\vartheta+\varphi_1)-i_{21}a\sin\gamma+p
\end{aligned}\right\}
$$

$$(4.15)$$

Knowing the conditions of relative velocity between worm and grinding wheel, the equation of contact can be written as:

$$
\vec{n}_{1F}\cdot\vec{v}_{1F}^{(12)} = 0 \tag{4.16}
$$

where $\vec{n}_{1F}$ is the normal vector of surface No 1.

Its components are:

$$
\left.\begin{aligned}
n_{1Fx} &= -\eta\sin\vartheta\frac{K-\eta}{\sqrt{\rho_{ax}^{2}-(K-\eta)^2}}+p\cdot\cos\vartheta \\
n_{1Fy} &= \eta\cos\vartheta\frac{K-\eta}{\sqrt{\rho_{ax}^{2}-(k-\eta)^2}}+p\cdot\sin\vartheta \\
n_{1Fz} &= \eta
\end{aligned}\right\} \tag{4.17}
$$

Substituting (4.15) and (4.17) into equation (4.16) the contact conditions can be obtained as:

$$
\left.\begin{aligned}
\vec{n}_{1F}\vec{v}_{1F}^{(12)} = 0 = \eta\,& \frac{K-\eta}{\sqrt{\rho_{ax}^2 -(K-\eta)^2}}\sin\gamma\cos(\vartheta+\varphi_1)z_1 + \\
+\eta a\,& \frac{K-\eta}{\sqrt{\rho_{ax}^2 -(K-\eta)^2}}\cos\gamma\sin(\vartheta+\varphi_1) - \\
-\eta p\cos\gamma &+ p\sin\gamma\sin(\vartheta+\varphi_1)z_1 - pa\cos\gamma\cos(\vartheta+\varphi_1) - \\
-\eta^2 \sin\gamma &\cos(\vartheta+\varphi_1) - \eta a\sin\gamma
\end{aligned}\right\} \quad (4.18)
$$

Equation (4.18) formulates the correlation between parameters η, ϑ of the contact point between the worm and grinding wheel and the angular displacement φ_1 of the worm (element No 1).

To determine the contact curve in coordinate system S_{1F} the following equation is used:

$$
\left.\begin{aligned}
\vec{n}_{1F}\vec{v}_{1F}^{(12)} &= 0 \\
x_{1F} &= x_{1F}(\eta,\vartheta) \\
y_{1F} &= y_{1F}(\eta,\vartheta) \\
z_{1F} &= z_{1F}(\eta,\vartheta)
\end{aligned}\right\} \quad (4.19)
$$

When movement parameter φ_1 is kept constant, equation (4.18) makes it possible to express one of the parameters and a vector–scalar function, belonging to the value φ_1, is obtained to formulate the equation of the contact curve.

The correlation between the parameters in explicit form cannot be determined for the equation of the contact curve when $\varphi_1 = $ constant, as is the situation in this case. By choosing discrete values for one of the parameters within the domain of validity belonging to the real tooth surface from equation (4.19), the coordinates of the contact curve can be calculated:

$$
\left.\begin{aligned}
\vec{n}_{1F}\cdot\vec{v}_{1F}^{(12)} &= 0 \\
\vec{r}_{1F} &= \vec{r}_{1F}(\eta,\vartheta) \\
\vec{r}_2 &= \underline{M}_{2,1F}\cdot\vec{r}_{1F}
\end{aligned}\right\} \quad (4.20)
$$

At the point of the characteristics rotated into the plane x_2, z_2 the axial section of the wheel is obtained (as in Figure 4.8).

The equation of the grinding wheel surface, shaped up as an enveloping surface of the series of worm surfaces, can be obtained in the S_{2F} system as:

$$\left.\begin{array}{c} \vec{n}_{1F} \cdot \vec{v}_{1F}^{(12)} = 0 \\[2mm] \vec{r}_{1F} = \vec{r}_{1F}(\eta,\vartheta) \\[2mm] \vec{r}_{2F} = \underline{M}_{2F,1F}\vec{r}_{1F} \end{array}\right\} \qquad (4.21)$$

The scalar equations equivalent to vector equation (4.21), expressing the loci of contact curves in the coordinate system fixed to the grinding wheel, are:

$$\left.\begin{array}{c} \vec{n}_{1F}\vec{v}_{1F}^{(12)} = 0 \\[3mm] x_{1F} = x_{1F}(\eta,\vartheta) \\ y_{1F} = y_{1F}(\eta,\vartheta) \\ z_{1F} = z_{1F}(\eta,\vartheta) \\[3mm] x_{2F} = x_{2F}(\eta,\vartheta,\varphi_1,\varphi_2) \\ y_{2F} = y_{2F}(\eta,\vartheta,\varphi_1,\varphi_2) \\ z_{2F} = z_{2F}(\eta,\vartheta,\varphi_1,\varphi_2) \end{array}\right\} \qquad (4.22)$$

A numerical example has been run with input data: $i = 12.5$, $a = 280$ mm, $m = 12$ mm, $\rho_{ax} = 50$ mm for grinding of the worm. The results of the calculations provided the points of the characteristics (equation (4.7) and Figure 4.12) and the axial section of the grinding wheels. The correlation between parameters in explicit form cannot be determined.

The coordinates of the grinding wheel profile calculated by the computer show very good correlation with the profile drawn by the profile projector from the profile manufactured by generation in practice.

The block diagram (Figure 3.17) describes the computer program of the calculations.

The input data were as shown in Figure 4.6.

Nomination	Symbol	Data		
Number of teeth worm	z_1	2	3	5
Module in axial section	m	2	12.5	16
Lead angle at reference cylinder	γ_0	12°5'45"	21°2'15"	27°45'30"
Direction of tooth inclination	left	left	Left	left
Radius of tooth arch in axial section	ρ_p	36.6	50	58.5
Addendum height to adjust at measurement of tooth chord thickness	$\overline{S}_{n1}$	9	10	15
Value of worm chord thickness	$\overline{f}_{n1}$	$10 \begin{smallmatrix} +0.0 \\ -0.125 \end{smallmatrix}$	$13 \begin{smallmatrix} +0.0 \\ -0.125 \end{smallmatrix}$	$13.44 \begin{smallmatrix} +0.0 \\ -0.125 \end{smallmatrix}$
Centre distance	a	280	280	280
Diameter of reference cylinder	d_{01}	84	97.5	152
Lead	H	56.5486	117.809722	251.327408
Profile angle in axial section	ρ_{ax}	22°28'30"	24°31'10"	23°9'
Number of teeth on worm gear	z_2	51	35	24
Tolerance of radial knock of worm teething	F_{r1}	±0.017		
Tolerance of axial pitch deviation of worm	f_{p1}	±0.016		
Tolerance of lead of thread	f_γ	±0.018		
Tolerance of worm profile inclination	f_f	0.08		

Figure 4.6 *Calculated and manufactured basic data of worms (wire-drawing machines) having curved profile*

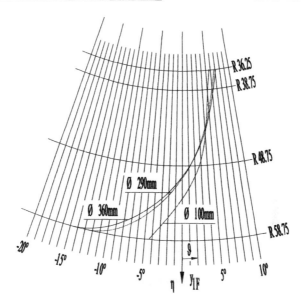

Figure 4.7 *Contact curves for a ⌀100 mm (relieved milling cutter) with grinding wheels having ⌀290 mm, ⌀350 mm, ⌀100 mm diameters for manufacturing curved profile worm*
(Data of worm–worm gear are: i = 12; a = 280 mm; P_1 = 40 kW; z_1 = 3; m = 12.5 mm; ρ_{ax} = 50 mm; K = 69.5 mm)

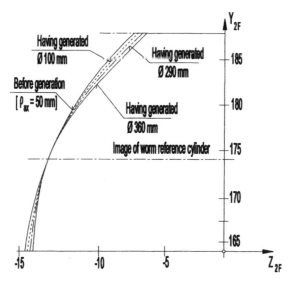

Figure 4.8 *Grinding wheel profiles belonging to contact curves shown in Figure 4.7*

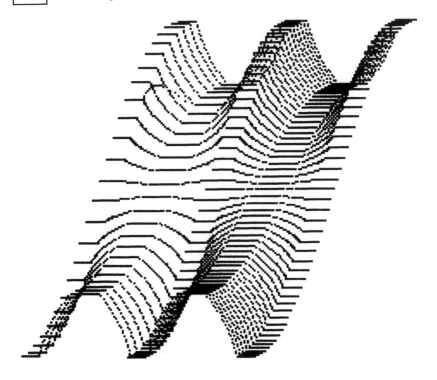

Figure 4.9 *Grinding of ZTA type worm having circular profile in axial section; contact curves*

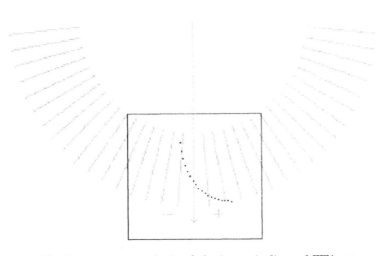

Figure 4.10 *Contact curves obtained during grinding of ZTA type worm, in polar coordinate system*

Geometrical data of worm		
Number of starting on worm		3.00
Modules	/mm/	12.00
Dedendum cylinder	/mm/	76.00
Reference cylinder diameter	/mm/	96.00
Addendum cylinder diameter	/mm/	116.00
Diameter of the generating cylinder	/mm/	69.00
Radius of curvature on profile	/mm/	50.00
Tooth thickness on reference cylinder	/mm/	13.00
Lead angle	/fok/	20.56
Profile angle	/fok/	24.83
Direction of lead	/j-b/	j
Diameter of grinding wheel	/mm/	290.00
Value of angular step	/0.2-3/	1.00
Allowable tolerance of root determination	/0.001-2/	0.001

$$z_{2F}$$

$$y_{2F}\,(R_K)$$

Contact curve of worm and grinding wheel in polar coordinate system		Dimensions of grinding wheel profile	
		z_{2F}	$y_{2F}(R_K)$
angular position/deg/	radius /mm/	Face width /mm/	radius /mm/
-12.00	57.951	14.035	127.610
-11.00	57.675	14.048	127.526
-10.00	57.355	14.050	127.512
-9.00	56.987	14.038	127.574
-8.00	56.563	14.010	127.720
-7.00	56.078	13.961	127.956
-6.00	55.521	13.886	128.291
-5.00	54.383	13.779	128.736
-4.00	54.154	13.632	129.301
-3.00	53.321	13.434	130.000
-2.00	52.369	13.173	130.845
-1.00	51.282	12.834	131.852
0.00	50.042	12.397	133.039
1.00	48.630	11.835	134.423
2.00	47.025	11.116	136.025
3.00	45.200	10.197	137.868
4.00	43.125	9.016	139.980
5.00	40.753	7.482	142.406

Figure 4.11 _Input-output data for the computer program determining machining grinding wheel profile of helicoidal surface having circular profile in axial section_

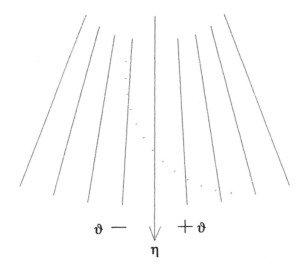

Figure 4.12 *Enlarged detail of Figure 4.10*

4.3 MACHINING GEOMETRY OF SPIROID DRIVES

In this section we review the results achieved by the use of CAD for tools for spiroid drives and their manufacture. The results obtained in the manufacture of conical helicoidal surfaces, especially in the machining of spiroid drives, have a double importance:

- the possibility to use conical surfaces as worm surfaces;
- the possibility to use conical surfaces as back surfaces for milling cutter teething (these can be radial or diagonal back surfaces).

The computer program, based on theoretical considerations and discussed in Section 3.2, provides the possibility to plot conical helicoid surfaces, to determine the contact curve between the conical helicoidal surface and the grinding wheel as well as to determine the grinding wheel profile. When determining the wheel profile, the possibility is open to optimize it, producing a mean profile variation of the grinding wheel using the algorithm displayed in Section 3.2.2.4. This result can only be utilized when regulating the wheel using a CNC-controlled wheel regulator but the knowledge of the position for optimal wheel generation can be utilized for mechanical wheel regulators too.

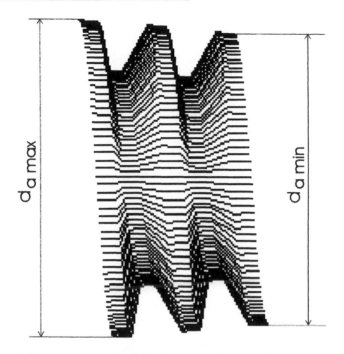

Figure 4.13 _KA type conical Archimedian helicoidal surface_
_(m = 5 mm; d_1 = 64 mm; β_j = 10°; β_b = 30°)_

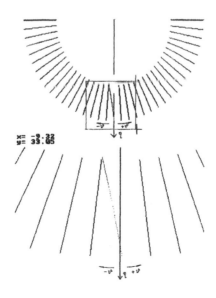

Figure 4.14 _Contact curve; grinding of KA type conical Archimedian worm,_
in polar coordinate system (minor diameter of worm $d_{a\,min}$)

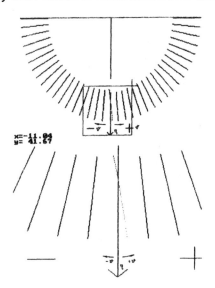

Figure 4.15 *Contact curve; grinding of KA type conical Archimedian worm, in polar coordinate system (major diameter of worm $d_{a\ max}$)*

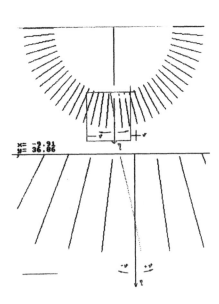

Figure 4.16 *Contact curve; grinding of KA type conical Archimedian worm, in polar coordinate system (mean diameter of worm $d_{a\ med}$)*

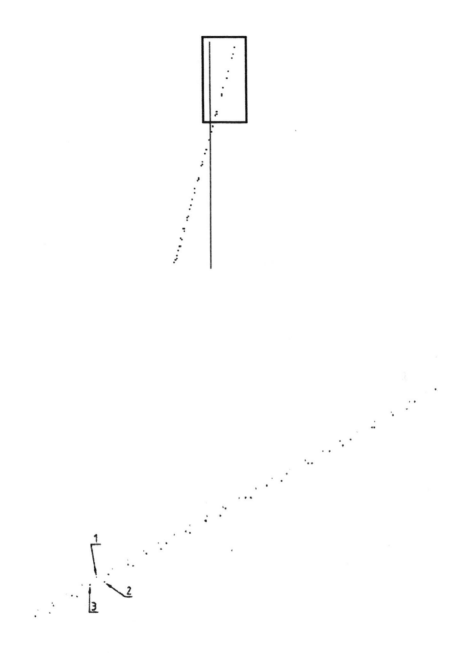

Figure 4.17 _Optimal tool profile required for grinding of KA type conical Archimedian worm_
1 Wheel profile for $d_{a\,min}$ 2 Wheel profile for $d_{a\,max}$ 3 Wheel profile for φ_{opt}.

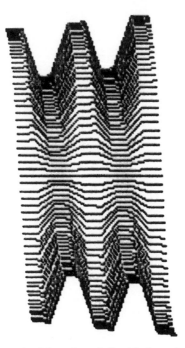

Figure 4.18 *KI type conical involute helicoidal surface (m = 5 mm; d_1 = 64 mm; z_1 = 1; β_j = 10°; β_b = 30°; r_D = 6.869 mm)*

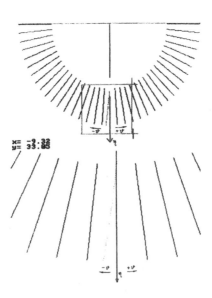

Figure 4.19 *Contact curve; grinding of KI type conical involute worm in polar coordinate system (minor diameter of worm $d_{a\,min}$)*

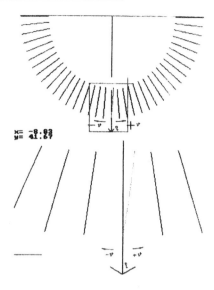

Figure 4.20 *Contact curve; grinding of KI type conical involute worm in polar coordinate system (major diameter of worm $d_{a\,max}$)*

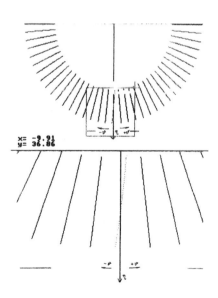

Figure 4.21 *Contact curve; grinding of KI type conical involute worm in polar coordinate system (mean diameter of worm $d_{a\,med}$)*

Z2	R2
0.216	74.927
0.361	74.349
0.520	73.713
0.698	73.003
0.901	72.196
1.137	71.251
1.427	70.096
1.813	68.555
2.490	65.859

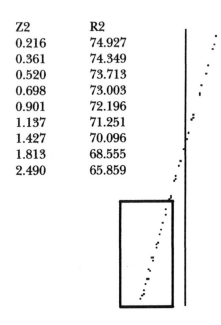

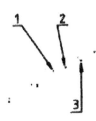

Figure 4.22 *Optimal tool profile required for grinding of KI type conical involute worm*
1 Wheel profile for $d_{a\ min}$ 2 Wheel profile for $d_{a\ max}$ 3 Wheel profile for φ_{opt}.

Figure 4.23 *KN2 type grooved convolute helicoidal surface*
(m = 5 mm; d_1 = 64 mm; z_1 = 1; β_j = 10°; β_b = 30°; r_D = 1.172 mm)

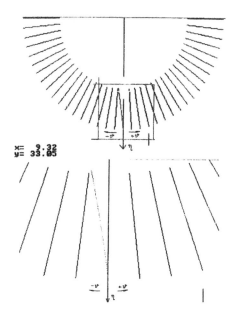

Figure 4.24 *Contact curve for grinding of KN2 type grooved convolute conical worm in polar coordinate system (minor diameter of worm $d_{a\,min}$)*

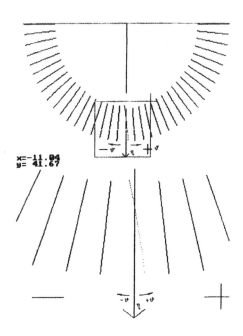

Figure 4.25 *Contact curve for grinding of KN2 type grooved convolute conical worm in polar coordinate system (major diameter of worm $d_{a\ max}$)*

Z2	R2
-0.224	74.811
-0.021	74.061
0.198	73.252
0.434	72.373
0.517	71.410
0.692	70.341
0.976	69.137
1.296	67.753
1.661	66.116
2.092	64.090
2.624	

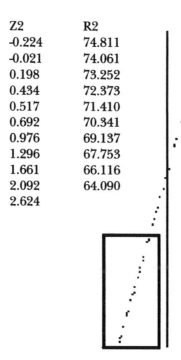

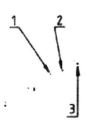

Figure 4.26 _Optimal tool profile required for grinding of KN2 type grooved conical convolute worm_

Figure 4.27 *KN1 type conical crawling convolute helicoidal surface* $(m = 5\ mm;\ d_1 = 64\ mm;\ z_1 = 1;\ \beta_j = 10^\circ;\ \beta_b = 30^\circ;\ r_D = 0.664\ mm)$

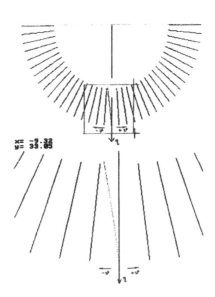

Figure 4.28 *Contact curve; grinding of KN1 type conical crawling convolute worm (minor diameter of worm $d_{a\ min}$)*

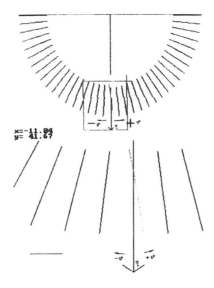

Figure 4.29 *Contact curve; grinding of KN1 type conical crawling convolute worm (major diameter of worm $d_{a\ max}$)*

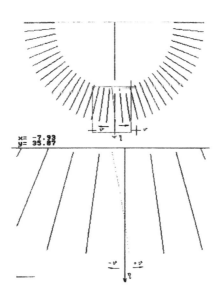

Figure 4.30 *Contact curve; grinding of KN1 type conical crawling convolute worm (mean diameter of worm $d_{a\ mean}$)*

Z2	R2
-0.323	74.985
-0.135	74.278
0.072	73.522
0.293	72.707
0.532	71.822
0.793	70.852
1.000	69.778
1.401	68.571
1.768	67.188
2.199	65.559

Figure 4.31 *Optimal tool profile required for grinding of KN1 type conical crawling convolute worm*
1 Wheel profile for $d_{a\ min}$ 2 Wheel profile for $d_{a\ max}$ 3 Wheel profile for φ_{opt}.

GEOMETRIC DATA OF WORM

		k (KA)
Type of worm		
N° of starting on worm	/mm/	1.00
Major diameter of worm	/mm/	74.00
Length of worm	/mm/	73.00
Tooth thickness on addendum cylinder in axial direction	/mm/	4.00
Tooth addendum height	/mm/	5.00
Tooth height	/mm/	11.00
Angle of side surface inclination on the left	/degree/	10.00
Angle of side surface inclination on the right	/degree/	30.00
Half cone angle of worm	/degree/	5.00
Diameter of grinding wheel	/mm/	150.00
Radius of base cylinder	/mm/	0.00
Direction of tooth inclination	/r-e/	r

GEOMETRIC SIZES OF GRINDING WHEEL PROFILE

z_2 face width /mm/	$R_2(y_2)$ radius /mm/	face width /mm/	radius /mm/
			Optimized profile
-0.407	74.904	-0.375	74.802
-0.333	74.643	-0.313	74.670
-0.258	74.373	-0.251	74.445
-0.181	74.204	-0.187	74.214
-0.102	73.825	-0.122	73.978
-0.023	73.538	-0.055	73.737
0.060	73.244	0.012	73.491
0.144	73.943	0.082	73.239
0.231	72.032	0.152	72.981
0.319	73.312	0.225	73.717
0.410	71.983	0.299	72.446
0.504	71.644	0.375	72.168
0.600	71.293	0.453	71.382
0.700	70.931	0.532	71.588
0.802	70.556	0.614	71.288
0.908	70.167	0.698	70.975
1.018	69.763	0.785	70.654
0.132	69.343	0.875	70.323
1.251	68.905	0.967	69.981
1.375	68.448	1.062	69.627
1.504	67.968	1.161	69.260
1.639	67.465	1.263	68.879
1.782	66.935	1.369	68.483
1.932	66.375	1.479	68.070
2.091	65.781	1.595	67.640
3.360	65.147	1.715	67.189
3.441	64.466	1.841	66.716
		1.973	66.218
-0.340	74.868	2.113	65.692
-0.221	74.425	2.261	65.135
-0.096	73.364		
0.034	73.480		
0.179	72.973		
0.313	72.440		
0.464	71.376		
0.623	71.379		
0.792	70.644		
0.973	69.964		
1.167	69.232		
1.378	68.438		
1.697	67.754		

$D_{a\ min}$ ↓ (upper left section)

$D_{a\ max}$ ↓ (lower left section)

Optimized profile ↓ (right section)

Figure 4.32 *Results of running program to determine grinding wheel profile in contact with spiroid worm coordinates obtained for $d_{a\ max}$ and $d_{a\ min}$; optimized profile can be obtained with $\varphi_{1\ opt.}$ (see Figure 4.33)*

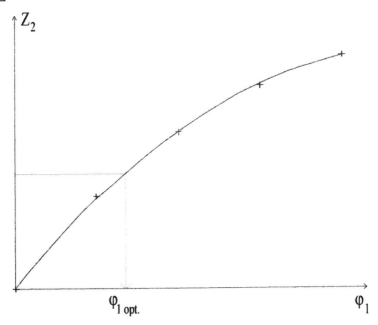

Figure 4.33 *Changes in the position, in direction z_2, for point on addendum cylinder of grinding wheel, as a function of movement parameter φ_1. The value of optimal movement parameter $\varphi_{1\,opt}$ corresponds to a halved variation in dimensions*

4.4 INTERSECTION OF CYLINDRICAL HELICOIDAL SURFACE HAVING CIRCULAR PROFILE IN AXIAL SECTION (ZTA) AND THE ARCHIMEDIAN THREAD FACE SURFACE AS GENERATING CURVE OF BACK SURFACE

An helicoid surface in axial section, having a circular profile representing the enveloping surface of a worm gear-generating milling cutter, can be described using a two-parameter equation system in the K_{1F} rotating coordinate system joined to the generating milling cutter. The equation system, using Figure 4.34, is:

$$
\left.
\begin{aligned}
x_{1F} &= -\eta \cdot \sin \vartheta \\
y_{1F} &= \eta \cdot \cos \vartheta \\
z_{1F} &= p_a \cdot \vartheta - \sqrt{\rho_{ax}^2 - (K - \eta)^2} + z_{axp}
\end{aligned}
\right\}
\tag{4.23}
$$

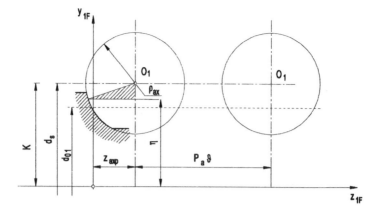

Figure 4.34 *Generation of cylindrical helicoid surface having circular profile in axial section*

where:

η, ϑ are the internal parameters of helicoidal surface,

ρ_{ax} is the radius of circular profile,

K, z_{axp} are coordinates of circular profile centre point for $\vartheta = 0$ in K_{1F} coordinate system,

d_s is the reference cylinder of milling cutter,

P_a is the axial thread parameter.

The equation of the Archimedian helicoidal surface in the K_{1F} coordinate system characterized by the perpendicular to the thread line on the reference cylinder of the wrapping surface is:

$$\left.\begin{aligned} x_{1F} &= -\eta \cdot \sin \vartheta \\ y_{1F} &= \eta \cdot \cos \vartheta \\ z_{1F} &= \frac{-r_s^2}{P_a} \cdot \vartheta \end{aligned}\right\} \tag{4.24}$$

The solution of the simultaneous equations for the two surfaces is mathematically the single parametric equation system of the edge curve $\vec{r}_e(\vartheta)$:

$$\left.\begin{aligned} x_{1F}(\vartheta) &= -\eta(\vartheta) \cdot \sin \vartheta \\ y_{1F}(\vartheta) &= \eta(\vartheta) \cdot \cos \vartheta \\ z_{1F}(\vartheta) &= \frac{-r_s^2}{P_a} \cdot \vartheta \end{aligned}\right\} \tag{4.25}$$

where:

$$\eta(\vartheta) = K \pm \sqrt{K^2 - \left[\frac{r_s^4}{p_a^2} \cdot \vartheta^2 + 2 \cdot r_s^2 \cdot \vartheta^2 + 2 \cdot \frac{r_s^2}{p_a} \cdot \vartheta \cdot z_{axp} + p_a^2 \cdot \vartheta^2 + 2 \cdot p_a \cdot \vartheta \cdot z_{axp} + z_{axp}^2 + K^2 - \rho_{ax}^2 \right]} \quad (4.26)$$

Only the negative sign can be interpreted in practice.

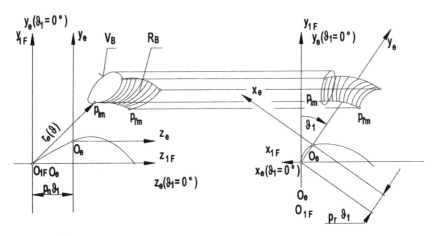

Figure 4.35 *Generation of the back surface on a generating milling cutter*

4.4.1 Generation of radial back surface with generator curve

The intersection curve $\bar{r}_e(\vartheta)$ created by generation of the back surface by radial relief is the generator (see Figure 4.35).

The parametric equations for coordinates of this curve can be obtained so that when the back surface parameter $\vartheta_1 = 0$, the coordinate systems K_{1F} and K_e (x_e, y_e, z_e) coincide.

The equations are:

$$\left. \begin{aligned} x_e(\vartheta) &= -\eta(\vartheta) \cdot \sin \vartheta \\ y_e(\vartheta) &= \eta(\vartheta) \cdot \cos \vartheta \\ z_e(\vartheta) &= -\frac{r_s^2}{p_a} \cdot \vartheta \end{aligned} \right\} \quad (4.27)$$

As in Figure 4.35, let the coordinate system K_e move, in the coordinate system K_{1F}, with a thread motion. Let the thread motion be determined by the parameters p_a (axial), p_r (radial) and ϑ_1 (angular).

During the relative displacement the curve, $\vec{r}_e$, determined by equations (4.27), remains in contact with the back surface $\vec{r}_{1F}(\vartheta,\vartheta_1)$.

The correlation between coordinate systems K_{1F} and K_e is determined by the matrix:

$$
M_{1F.e} = \begin{bmatrix} \cos\vartheta_1 & -\sin\vartheta_1 & 0 & p_r\cdot\vartheta_1\cdot\cos\vartheta_1 \\ \sin\vartheta_1 & \cos\vartheta_1 & 0 & p_r\cdot\vartheta_1\cdot\sin\vartheta_1 \\ 0 & 0 & 1 & p_a\cdot\vartheta_1 \\ 0 & 0 & 0 & 1 \end{bmatrix} \tag{4.28}
$$

Using the matrix $\underline{M}_{1F.e}$ the equation of the back surface is:

$$
\vec{r}_{1F}(\vartheta,\vartheta_1) = \underline{M}_{1F.e}\cdot\vec{r}_e(\vartheta) =
$$

$$
= \begin{bmatrix} -\eta(\vartheta)\cdot\sin\vartheta\cdot\cos\vartheta_1 - \eta(\vartheta)\cdot\cos\vartheta\cdot\sin\vartheta_1 + p_r\cdot\vartheta_1\cdot\cos\vartheta_1 \\ -\eta(\vartheta)\cdot\sin\vartheta\cdot\sin\vartheta_1 - \eta(\vartheta)\cdot\cos\vartheta\cdot\cos\vartheta_1 + p_r\cdot\vartheta_1\cdot\sin\vartheta_1 \\ \dfrac{r_s^2}{p_a}\cdot\vartheta + p_a\cdot\vartheta_1 \\ 1 \end{bmatrix}. \tag{4.29}
$$

4.4.2 Contact curve of the back surface and the grinding wheel

As the back surface is a conical helicoidal surface, the kinematic arrangement for machining is the same as shown in Figure 3.27.

This model derived from the general model (Figure 4.1), as reviewed in (4.3), can be derived by substituting $\alpha = 0°$ and $c = 0$. Using the same method, the relative velocity vector between the two bodies can be determined by deriving the matrix $\underline{P}_{1s}$ from the general matrix $\underline{P}_{1k}$ by applying equations (3.78) and (3.86), ie:

$$
\vec{v}_{1F}^{(1,2)} = \underline{P}_{1a}\cdot\vec{r}_{1F} = \begin{bmatrix} v_{1Fx} \\ v_{1Fy} \\ v_{1Fz} \end{bmatrix} \tag{4.30}
$$

For the sake of simplicity let η be written instead of $\eta(\vartheta)$ only for the components that can be obtained from the following equations:

$$
\begin{aligned}
v_{1Fx} =\ & i \cdot \eta \cdot \cos\delta \cdot \cos\gamma \cdot \sin\vartheta \cdot \sin\vartheta_1 - i \cdot \eta \cdot \cos\delta \cdot \cos\gamma \cdot \cos\vartheta \cdot \cos\vartheta_1 - \\
& - i \cdot p_r \cdot \cos\delta \cdot \cos\gamma \cdot \vartheta_1 \cdot \sin\vartheta_1 + \eta \cdot \sin\vartheta \cdot \sin\vartheta_1 - \eta \cdot \cos\vartheta \cdot \cos\vartheta_1 - p_r \cdot \vartheta_1 \cdot \sin\vartheta_1 + \\
& + i_{21} \cdot \frac{r_s^2}{p_a} \cdot \cos\varphi_1 \cdot \sin\delta \cdot \cos\gamma \cdot \vartheta - i \cdot p_a \cdot \cos\varphi_1 \cdot \sin\delta \cdot \cos\gamma \cdot \vartheta_1 - i \cdot \frac{r_s^2}{p_a} \cdot \sin\varphi_1 \cdot \sin\gamma \cdot \vartheta + \\
& + i \cdot p_a \cdot \sin\varphi_1 \cdot \sin\gamma \cdot \vartheta_1 + i \cdot \varphi_1 \cdot p_r \cdot \sin\varphi_1 \cdot \sin\delta \cdot \cos\delta \cdot \sin\gamma + i \cdot \varphi_1 \cdot p_a \cdot \sin\varphi_1 \cdot \cos^2\delta \cdot \sin\gamma - \\
& - i \cdot z_{ax} \cdot \cos\varphi_1 \cdot \sin\delta \cdot \cos\gamma - i \cdot a \cdot \cos\varphi_1 \cdot \cos\gamma + i \cdot z_{ax} \cdot \sin\varphi_1 \cdot \sin\gamma - p_r \cdot \sin\varphi_1 \cdot \sin^2\delta - \\
& - p_a \cdot \sin\varphi_1 \cdot \sin\delta \cdot \cos\delta + i \cdot a \cdot \sin\varphi_1 \cdot \sin\delta \cdot \sin\gamma \\[1.2em]
v_{1Fy} =\ & -i \cdot \eta \cdot \cos\delta \cdot \cos\gamma \cdot \sin\vartheta \cdot \sin\vartheta_1 - i \cdot \eta \cdot \cos\delta \cdot \cos\gamma \cdot \cos\vartheta \cdot \sin\vartheta_1 - \\
& i \cdot p_r \cdot \cos\delta \cdot \cos\gamma \cdot \vartheta_1 \cdot \cos\vartheta_1 - \eta \cdot \sin\vartheta \cdot \cos\vartheta_1 - \eta \cdot \cos\vartheta \cdot \sin\vartheta_1 + p_r \cdot \vartheta_1 \cdot \cos\vartheta_1 - \\
& - i \cdot \frac{r_s^2}{p_a} \cdot \sin\varphi_1 \cdot \sin\delta \cdot \cos\gamma \cdot \vartheta + i \cdot p_a \cdot \sin\varphi_1 \cdot \sin\delta \cdot \cos\gamma \cdot \vartheta_1 - i \cdot \frac{r_s^2}{p_a} \cdot \cos\varphi_1 \cdot \sin\gamma \cdot \vartheta + \\
& + i \cdot p_a \cdot \cos\varphi_1 \cdot \sin\gamma \cdot \vartheta_1 + i \cdot \varphi_1 \cdot p_r \cdot \cos\varphi_1 \cdot \sin\delta \cdot \cos\delta \cdot \sin\gamma + i \cdot \varphi_1 \cdot p_a \cdot \cos\varphi_1 \cdot \cos^2\delta \cdot \sin\gamma + \\
& + i \cdot z_{ax} \cdot \sin\varphi_1 \cdot \sin\delta \cdot \cos\gamma + i \cdot a \cdot \sin\varphi_1 \cdot \cos\gamma + i \cdot z_{ax} \cdot \cos\varphi_1 \cdot \sin\gamma - p_r \cdot \cos\varphi_1 \cdot \sin^2\delta - \\
& - p_a \cdot \cos\varphi_1 \cdot \sin\delta \cdot \cos\delta + i \cdot a \cdot \cos\varphi_1 \cdot \sin\delta \cdot \sin\gamma \\[1.2em]
v_{1Fz} =\ & i \cdot \eta \cdot \sin\varphi_1 \cdot \sin\gamma \cdot \sin\vartheta \cdot \cos\vartheta_1 + i \cdot \eta \cdot \sin\varphi_1 \cdot \sin\gamma \cdot \cos\vartheta \cdot \sin\vartheta_1 - \\
& - i \cdot p_r \cdot \sin\varphi_1 \cdot \sin\gamma \cdot \vartheta_1 \cdot \cos\vartheta_1 - i \cdot \eta \cdot \cos\varphi_1 \cdot \sin\delta \cdot \cos\gamma \cdot \sin\vartheta \cdot \cos\vartheta_1 - \\
& - i \cdot \eta \cdot \cos\varphi_1 \cdot \sin\delta \cdot \cos\gamma \cdot \cos\vartheta \cdot \sin\vartheta_1 + i \cdot p_r \cdot \cos\varphi_1 \cdot \sin\delta \cdot \cos\gamma \cdot \vartheta_1 \cos\vartheta_1 + \\
& + i \cdot \eta \cdot \cos\varphi_1 \cdot \sin\gamma \cdot \sin\vartheta \cdot \sin\vartheta_1 - i \cdot \eta \cdot \cos\varphi_1 \cdot \sin\gamma \cdot \cos\vartheta \cdot \cos\vartheta_1 - \\
& - i \cdot p_r \cdot \cos\varphi_1 \cdot \sin\gamma \cdot \vartheta_1 \cdot \sin\vartheta_1 + i \cdot \eta \cdot \sin\varphi_1 \cdot \sin\delta \cdot \cos\gamma \cdot \sin\vartheta \cdot \sin\vartheta_1 - \\
& - i \cdot \eta \cdot \sin\varphi_1 \cdot \sin\delta \cdot \cos\gamma \cdot \cos\vartheta \cdot \cos\vartheta_1 - i \cdot p_r \cdot \sin\varphi_1 \cdot \sin\delta \cdot \cos\gamma \cdot \vartheta_1 \cdot \sin\vartheta_1 + \\
& + i \cdot \varphi_1 \cdot p_r \cdot \sin^2\delta \cdot \sin\gamma + p_r \cdot \sin\delta \cdot \cos\delta + p_a \cdot \cos^2\delta - \\
& - i \cdot a \cdot \cos\delta \cdot \sin\gamma + i \cdot \varphi_1 \cdot p_a \cdot \sin\delta \cdot \cos\delta \cdot \sin\gamma
\end{aligned}
\right\} \quad (4.31)
$$

The normal vector of the back surface is:

$$
\vec{n}_{1F} = \frac{\partial \vec{r}_{1F}}{\partial \vartheta} \times \frac{\partial \vec{r}_{1F}}{\partial \vartheta_1} \tag{4.32}
$$

After completing partial differentiation and calculating the vector products for the components of the normal vector:

$$
\begin{aligned}
n_{1Fx} =\ & -p_a \cdot \frac{\partial\eta}{\partial\vartheta} \cdot \sin\vartheta \cdot \sin\vartheta_1 - p_a \cdot \eta \cdot \cos\vartheta \cdot \sin\vartheta_1 + p_a \cdot \frac{\partial\eta}{\partial\vartheta} \cdot \cos\vartheta \cdot \cos\vartheta_1 - \\
& - p_a \cdot \eta \cdot \sin\vartheta \cdot \cos\vartheta_1 - \frac{r_s^2}{p_a} \cdot \eta \cdot \sin\vartheta \cdot \cos\vartheta_1 - \frac{r_s^2}{p_a} \cdot \eta \cdot \cos\vartheta \cdot \sin\vartheta_1 + \\
& + \frac{r_s^2}{p_a} \cdot p_r \cdot \sin\vartheta_1 + \frac{r_s^2}{p_a} \cdot p_r \cdot \vartheta_1 \cdot \cos\vartheta_1 \\[1.2em]
n_{1Fy} =\ & p_a \cdot \frac{\partial\eta}{\partial\vartheta} \cdot \sin\vartheta \cdot \cos\vartheta_1 + p_a \cdot \eta \cdot \cos\vartheta \cdot \cos\vartheta_1 - p_a \cdot \frac{\partial\eta}{\partial\vartheta} \cdot \cos\vartheta \cdot \sin\vartheta_1 - \\
& - p_a \cdot \eta \cdot \sin\vartheta \cdot \sin\vartheta_1 - \frac{r_s^2}{p_a} \cdot \eta \cdot \sin\vartheta \cdot \sin\vartheta_1 + \frac{r_s^2}{p_a} \cdot \eta \cdot \cos\vartheta \cdot \cos\vartheta_1 - \\
& - \frac{r_s^2}{p_a} \cdot p_r \cdot \cos\vartheta_1 + \frac{r_s^2}{p_a} \cdot p_r \cdot \vartheta_1 \cdot \sin\vartheta_1 \\[1.2em]
n_{1Fz} =\ & \eta \cdot \frac{\partial\eta}{\partial\vartheta} - p_r \cdot \frac{\partial\eta}{\partial\vartheta} \cdot \vartheta_1 \cdot \sin\vartheta_1 - p_r \cdot \frac{\partial\eta}{\partial\vartheta} \cdot \cos\vartheta + \eta \cdot p_r \cdot \sin\vartheta - \eta \cdot p_r \cdot \vartheta_1 \cdot \cos\vartheta
\end{aligned}
\right\} \quad (4.33)
$$

where:

$$\frac{\partial \eta}{\partial \vartheta} = \frac{-\left[\dfrac{r_s^4}{p_a^2} \cdot \vartheta + 2 \cdot r_s^2 \cdot \vartheta + \dfrac{r_s^2}{p_a} \cdot z_{axp} + p_a^2 \cdot \vartheta + p_a \cdot z_{axp}\right]}{\sqrt{-\dfrac{r_s^4}{p_a^2} \cdot \vartheta^2 - 2 \cdot r_s^2 \cdot \vartheta^2 - \dfrac{2 \cdot r_s^2}{p_a} \cdot \vartheta \cdot z_{axp} - p_a^2 \cdot \vartheta^2 - 2 \cdot p_a \cdot \vartheta \cdot z_{axp} - z_{axp}^2 + \rho_{ax}^2}}$$

From equations (4.31) and (4.33) the equation for contact between the points of the back surface and grinding wheel can be written as:

$$\vec{n}_{1F} \cdot \vec{v}_{1F}^{(12)} = 0$$

$$= (i \cdot \cos\delta \cdot \cos\gamma + 1) \cdot (\eta \cdot \frac{1}{p_a} \cdot p_r \cdot r_s^2 \cdot \sin\vartheta \cdot \frac{\partial \eta}{\partial \vartheta} \cdot \eta \cdot p_a) +$$

$$+ (p_a \cdot \vartheta_1 \cdot \frac{r_s^2}{p_a} \cdot \vartheta) \cdot (\frac{\partial \eta}{\partial \vartheta} \cdot i \cdot p_a \cdot \cos\gamma \cdot \sin\delta \cdot \sin\vartheta_1 \cdot \sin\theta + \frac{\partial \eta}{\partial \vartheta} \cdot i \cdot p_a \cdot \sin\gamma \cdot \sin(\vartheta_1 + \theta) +$$

$$+ i \cdot \eta \cdot p_a \cdot \cos\gamma \cdot \sin\delta \cdot \sin(\vartheta_1 + \theta) + i \cdot \eta \cdot p_a \cdot \sin\gamma \cdot \cos\vartheta_1 \cdot \cos\theta +$$

$$+ i \cdot \eta \cdot \frac{1}{p_a} \cdot r_s^2 \cdot \cos\gamma \cdot \sin\delta \cdot \sin(\vartheta_1 + \theta) + i \cdot \eta \cdot \frac{1}{p_a} \cdot r_s^2 \cdot \sin\gamma \cdot \cos\vartheta_1 \cdot \cos\theta -$$

$$- i \cdot \eta \cdot p_a \cdot \sin\gamma \cdot \sin\vartheta_1 \cdot \sin\theta - \frac{\partial \eta}{\partial \vartheta} \cdot i \cdot p_a \cdot \cos\gamma \cdot \sin\delta \cdot \cos\vartheta_1 \cdot \cos\theta -$$

$$- i \cdot \eta \cdot \frac{1}{p_a} \cdot r_s^2 \cdot \sin\gamma \cdot \sin\vartheta_1 \cdot \sin\theta - i \cdot \frac{1}{p_a} \cdot p_r \cdot r_s^2 \cdot \cos\gamma \cdot \sin\gamma \cdot \sin(\vartheta_1 + \varphi_1) -$$

$$- i \cdot \frac{1}{p_a} \cdot p_r \cdot r_s^2 \cdot \sin\gamma \cdot \cos(\vartheta_1 + \varphi_1)) + (z + a \cdot \sin\delta) \cdot (\frac{\partial \eta}{\partial \vartheta} \cdot i \cdot p_a \cdot \sin\gamma \cdot \sin(\vartheta_1 + \theta) +$$

$$+ i \cdot \eta \cdot p_a \cdot \sin\gamma \cdot \cos\vartheta_1 \cdot \cos\theta + i \cdot \eta \cdot \frac{1}{p_a} \cdot r_s^2 \cdot \sin\gamma \cdot \cos\vartheta_1 \cdot \cos\theta +$$

$$+ i \cdot \frac{1}{p_a} \cdot p_r \cdot r_s^2 \cdot \vartheta_1 \cdot \sin\gamma \cdot \sin(\vartheta_1 + \varphi_1) - i \cdot \eta \cdot p_a \cdot \sin\gamma \cdot \sin\vartheta_1 \cdot \sin\theta -$$

$$- i \cdot \eta \cdot \frac{1}{p_a} \cdot r_s^2 \cdot \sin\gamma \cdot \sin\vartheta_1 \cdot \sin\theta - i \cdot \frac{1}{p_a} \cdot p_r \cdot r_s^2 \cdot \sin\gamma \cdot \cos(\vartheta_1 + \varphi_1)) +$$

$$+ (z \cdot \sin\delta + a) \cdot (\frac{\partial \eta}{\partial \vartheta} \cdot i \cdot p_a \cdot \cos\gamma \cdot \sin\vartheta_1 \cdot \sin\theta + i \cdot \eta \cdot p_a \cdot \cos\gamma \cdot \sin(\vartheta_1 + \theta) +$$

$$+ i \cdot \eta \cdot \frac{1}{p_a} \cdot r_s^2 \cdot \cos\gamma \cdot \sin(\vartheta_1 + \theta) - \frac{\partial \eta}{\partial \vartheta} \cdot i \cdot p_a \cdot \cos\gamma \cdot \cos\vartheta_1 \cdot \cos\theta -$$

$$- i \cdot \frac{1}{p_a} \cdot p_r \cdot r_s^2 \cdot \cos\gamma \cdot \sin(\vartheta_1 + \varphi_1) - i \cdot \frac{1}{p_a} \cdot p_r \cdot r_s^2 \cdot \vartheta_1 \cdot \cos\gamma \cdot \cos(\vartheta_1 + \varphi_1)) +$$

$$+ (p_r \cdot \sin\delta + p_a \cdot \cos\delta) \cdot (\frac{\partial \eta}{\partial \vartheta} \cdot i \cdot p_a \cdot \varphi_1 \cdot \sin\gamma \cdot \cos\delta \cdot \sin(\vartheta_1 + \theta) +$$

$$+ \eta \cdot p_a \cdot \sin\delta \cdot \sin\vartheta_1 \cdot \sin\theta + i \cdot \eta \cdot p_a \cdot \varphi_1 \cdot \sin\gamma \cdot \cos\delta \cdot \cos\vartheta_1 \cdot \cos\theta +$$

$$+ i \cdot \eta \cdot \frac{1}{p_a} \cdot r_s^2 \cdot \varphi_1 \cdot \sin\gamma \cdot \cos\delta \cdot \cos\vartheta_1 \cdot \cos\theta + \eta \cdot \frac{1}{p_a} \cdot r_s^2 \cdot \sin\delta \cdot \sin\vartheta_1 \cdot \sin\theta +$$

(4.34)

$$+\frac{1}{p_a}\cdot p_r\cdot r_s^2\cdot\sin\delta\cdot\cos(\vartheta_1+\varphi_1)+i\cdot\frac{1}{p_a}\cdot p_r\cdot r_s^2\cdot\vartheta_1\cdot\varphi_1\cdot\sin\gamma\cdot\cos\delta\cdot\sin(\vartheta_1+\varphi_1)+$$

$$+i\cdot\eta\cdot p_r\cdot\varphi_1\cdot\sin\gamma\cdot\sin\delta\cdot\sin\vartheta+\eta\cdot p_r\cdot\cos\delta\cdot\sin\vartheta+$$

$$+\frac{\partial\eta}{\partial\vartheta}\cdot i\cdot\eta\cdot\varphi_1\cdot\sin\gamma\cdot\sin\delta+\frac{\partial\eta}{\partial\vartheta}\cdot\eta\cdot\cos\delta-\frac{\partial\eta}{\partial\vartheta}\cdot p_a\cdot\sin\delta\cdot\sin(\vartheta_1+\theta)-$$

$$-i\cdot\eta\cdot p_a\cdot\varphi_1\cdot\sin\gamma\cdot\cos\delta\cdot\sin\vartheta_1\cdot\sin\theta-\eta\cdot p_a\cdot\sin\delta\cdot\cos\vartheta_1\cdot\cos\theta-$$

$$-\eta\cdot\frac{1}{p_a}\cdot r_s^2\cdot\sin\delta\cdot\cos\vartheta_1\cdot\cos\theta-i\cdot\eta\cdot\frac{1}{p_a}\cdot r_s^2\cdot\varphi_1\cdot\sin\gamma\cdot\cos\delta\cdot\sin\vartheta_1\cdot\sin\theta-$$

$$-i\cdot\frac{1}{p_a}\cdot p_r\cdot r_s^2\cdot\varphi_1\cdot\sin\gamma\cdot\cos\delta\cdot\cos(\vartheta_1+\varphi_1)-\frac{1}{p_a}\cdot p_r\cdot r_s^2\cdot\vartheta_1\cdot\sin\delta\cdot\sin(\vartheta_1+\varphi_1)-$$

$$-\frac{\partial\eta}{\partial\vartheta}\cdot i\cdot p_r\cdot\varphi_1\cdot\sin\gamma\cdot\sin\delta\cdot\cos\vartheta-\frac{\partial\eta}{\partial\vartheta}\cdot p_r\cdot\cos\delta\cdot\cos\vartheta-$$

$$-\frac{\partial\eta}{\partial\vartheta}\cdot i\cdot p_r\cdot\vartheta_1\cdot\varphi_1\cdot\sin\gamma\cdot\sin\delta\cdot\sin\vartheta-\frac{\partial\eta}{\partial\vartheta}\cdot p_r\cdot\vartheta_1\cdot\cos\delta\cdot\sin\vartheta-$$

$$-i\cdot\eta\cdot p_r\cdot\vartheta_1\cdot\varphi_1\cdot\sin\gamma\cdot\sin\delta\cdot\cos\vartheta-\eta\cdot p_r\cdot\vartheta_1\cdot\cos\delta\cdot\cos\vartheta+$$

$$+(\eta\cdot p_r\cdot\sin\vartheta+\frac{\partial\eta}{\partial\vartheta}\cdot\eta)\cdot(i\cdot\eta\cdot\sin\gamma\cdot\sin\vartheta_1\cdot\sin\theta+$$

$$+i\cdot p_r\cdot\vartheta_1\cdot\cos\gamma\cdot\sin\delta\cdot\cos(\vartheta_1+\varphi_1)-i\cdot\eta\cdot\sin\gamma\cdot\cos\vartheta_1\cdot\cos\theta-$$

$$-i\cdot p_r\cdot\vartheta_1\cdot\sin\gamma\cdot\sin(\vartheta_1+\varphi_1)-i\cdot\eta\cdot\cos\gamma\cdot\sin\delta\cdot\sin(\theta+\vartheta_1)-$$

$$-i\cdot a\cdot\sin\gamma\cdot\cos\delta+$$

$$\left.\begin{array}{l}\end{array}\right\}\quad(4.34)$$

$$+(\frac{\partial\eta}{\partial\vartheta}\cdot p_r\cdot\vartheta_1\cdot\sin\vartheta+\eta\cdot p_r\cdot\vartheta_1\cdot\cos\vartheta)\cdot(i\cdot p_a\cdot\cos\gamma\cdot\cos\delta+$$

$$+i\cdot\eta\cdot\sin\gamma\cdot\cos\vartheta_1\cdot\cos\theta-i\cdot p_r\cdot\vartheta_1\cdot\sin\gamma\cdot\sin(\vartheta_1+\varphi_1)+$$

$$+i\cdot\eta\cdot\cos\gamma\cdot\sin\delta\cdot\sin(\theta+\vartheta_1)+i\cdot a\cdot\sin\gamma\cdot\cos\delta-$$

$$-i\cdot\eta\cdot\sin\gamma\cdot\sin\vartheta_1\cdot\sin\theta-i\cdot p_r\cdot\vartheta_1\cdot\cos\gamma\cdot\sin\delta\cdot\cos(\vartheta_1+\varphi_1)+$$

$$+(r_s^2\cdot\vartheta_1+\frac{\partial\eta}{\partial\vartheta}\cdot p_r\cdot\cos\theta)\cdot(i\cdot p_r\cdot\vartheta_1\cdot\sin\gamma)\cdot\sin(\vartheta_1+\varphi_1)-$$

$$-i\cdot p_r\cdot\vartheta_1\cdot\cos\gamma\cdot\sin\delta\cdot\cos(\vartheta_1+\varphi_1))+\frac{\partial\eta}{\partial\vartheta}\cdot i\cdot\eta\cdot p_r\cdot\sin\gamma\cdot\cos\vartheta\cdot\cos(\vartheta_1+\theta)+$$

$$+(\cos\gamma\cdot\sin\delta\cdot\cos(\vartheta_1+\varphi_1)-\sin\gamma\cdot\sin(\vartheta_1+\varphi_1))\cdot i\cdot\frac{1}{p_a^2}\cdot p_r\cdot r_s^4\cdot\vartheta_1\cdot\vartheta+$$

$$+(\vartheta_1\cdot\cos\vartheta_1-p_r\cdot\sin\vartheta_1)\cdot(\frac{1}{p_a}\cdot p_r\cdot r_s^2\cdot\vartheta_1\cdot\sin\vartheta_1)+$$

$$+(1-p_r)\cdot(\eta\cdot\frac{1}{p_a}\cdot r_s^2\cdot\vartheta_1\cdot\cos\vartheta_1\cdot\cos(\vartheta+\vartheta_1)+$$

$$+p_a\cdot\vartheta_1\cdot(\eta\cdot p_r\cdot\sin\vartheta_1\cdot\sin(\vartheta+\vartheta_1)+\frac{\partial\eta}{\partial\vartheta}\cdot\cos\vartheta_1\cdot\sin(\vartheta+\vartheta_1)+$$

$$+\eta\cdot\cos\vartheta_1\cdot\cos(\vartheta+\vartheta_1)-\frac{\partial\eta}{\partial\vartheta}\cdot p_r\cdot\sin\vartheta_1\cdot\cos(\vartheta+\vartheta_1))+$$

$$+\frac{1}{p_a}\cdot p_r\cdot r_s^2\cdot\vartheta_1\cdot(-i\cdot p_r\cdot\cos\gamma\cdot\cos\delta-p_r\cdot\vartheta_1\cdot\sin\vartheta_1\cdot\cos\vartheta_1-\cos^2\vartheta_1)+$$

$$+\frac{\partial\eta}{\partial\vartheta}\cdot i\cdot p_r\cdot\cos\vartheta\cdot(\eta\cdot\cos\gamma\cdot\sin\delta\cdot\sin(\theta+\varphi_1)+a\cdot\sin\gamma\cdot\cos\delta)$$

where $\theta = \varphi_1 + \vartheta$.

From the point of view of the machining geometry, a back surface, shaped by radial relief, can be inserted into conical helicoid surfaces. In other words, using the general kinematic model, the mated tool profile necessary for its meshing can be determined as has been seen in this chapter. Cylindrical or conical helicoidal surfaces and back surfaces machined by axial, radial or diagonal relief can all be handled using this model.

The tools manufactured will be discussed in Section 4.5.

4.5 MANUFACTURED TOOLS FOR WORM GEAR GENERATION AND OTHER TOOLS HAVING HELICOIDAL SURFACES

4.5.1 Design and manufacture of worm gear milling cutters

The practical realization of a theoretically precise worm milling cutter creates serious technological difficulties and, consequently, approximate solutions are chosen to keep the deviations within the tolerance range.

Objectives:

- to increase the service life of the tool, the possibility for resharpening should be increased (Figure 4.42a, b);
- to develop a method of relief suitable for the precise shaping of generator milling cutters using traditional devices, the tools fulfilling practical requirements.

Because back surfaces can be shaped up either as cylindrical or conical helicoidal surfaces, no new theoretical approach is necessary to machine them. Determining the equation of the back surface using the method presented in Section 3.3, then using this equation, a mated grinding wheel profile can be derived that has practically the same wheel profiling as the profile calculated for the task to be solved. To carry out this task there are several possibilities:

- to cut out a master template, according to the calculated profile, using a CNC-controlled electric-sparking machine having wire electrodes, to copy precisely the wheel-profile;

- applying a CNC-controlled wheel-regulator device. The author has developed such a device, licensing it as an invention of duty, to solve this problem (Dudás, 1988a).

The profile corrections carried out using the device gave good results regarding the precision of the grinding wheel profile.

Relief both of the flyblade and the milling cutter can be carried out using finger-shaped, disc-shaped or a hollow cylindrical-shaped grinding wheel (see Figures 4.40 and 4.41). The diameter of the grinding wheel in the case of the disc-shaped or hollow cylindrical-shaped tool can be determined knowing the value of the relief and the geometry of the milling cutter.

Investigating the axial profiles of tools manufactured in this way, it was found that they fulfilled the requirements of precision in shape (Figure 4.42).

Naturally, in the case of radial relief after resharpening of the tool, a deviation in radial size occurs (transfer of cutting edge, corrections Δa, $\Delta \gamma$ are necessary as Figure 4.38b shows). The effect of this phenomenon was investigated in the context of possible resharpening (Dudás, 1998e). The calculated sizes measured by profile gauge in the axial sections of the tool show good correlation. Significant defaults in profile, owing to changes in radial size, have not been detected. Regarding the normal resharpening reserve [$(0.05 \div 0.1) \times$ module], the value of default remained less than the allowable profile distortion (Figure 4.42) for this new manufacturing method. The contact area given in Figure 8.4 testifies to this fact. Figure 8.9 was obtained during manufacture of the worm gear after removing the complete resharpening reserve.

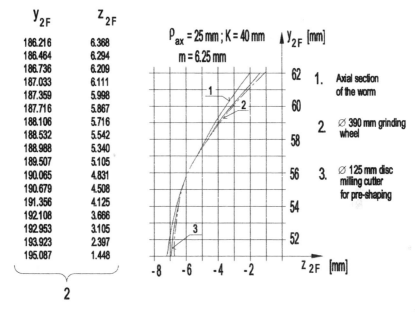

y_{2F}	z_{2F}
186.216	6.368
186.464	6.294
186.736	6.209
187.033	6.111
187.359	5.998
187.716	5.867
188.106	5.716
188.532	5.542
188.988	5.340
189.507	5.105
190.065	4.831
190.679	4.508
191.356	4.125
192.108	3.666
192.953	3.105
193.923	2.397
195.087	1.448

Figure 4.36 _Wheel profiles necessary to grind arched profile worm, as a function of diameter (for power transmission OL-3 type metal sheet cutter)_

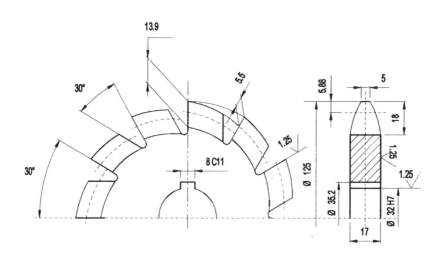

Figure 4.37 _Profile of circular milling cutter for worm having circular profile in axial section_
Data of worm: M = 6.25 mm; K = 40 mm; ρ_{ax} = 25 mm (Dudás, 1986e)

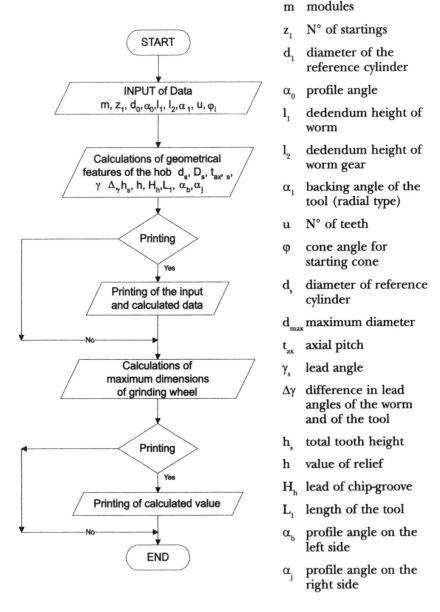

m   modules

z_1 N° of startings

d_1 diameter of the reference cylinder

α_0 profile angle

l_1 dedendum height of worm

l_2 dedendum height of worm gear

α_1 backing angle of the tool (radial type)

u   N° of teeth

φ cone angle for starting cone

d_s diameter of reference cylinder

d_{max} maximum diameter

t_{ax} axial pitch

γ_s lead angle

$\Delta\gamma$ difference in lead angles of the worm and of the tool

h_s total tooth height

h   value of relief

H_h lead of chip-groove

L_1 length of the tool

α_b profile angle on the left side

α_j profile angle on the right side

Figure 4.38 *Block diagram to determine geometric data of the tool, generator milling cutter for worm gear*

Design of ZA type worm milling cutter				
Basic data				
Modules	/mm/:		4.5000	
N° of starts	/mm/:		1.0000	
Diameter of the reference cylinder	/mm/:		72.6000	
Lead angle of the worm	/grade/:		3.5469	3°32`48``
Angle of the tooth side	/grade/:		20.0000	
Dedendum height of the worm tooth	/mm/:		5.4000	
Dedendum height of the worm gear	/mm/:		5.4000	
Angle of the back surface on the tool	8-10 /grade/:		8.0000	
Chosen angle of the back surface	/pcs/:		10.0000	
Angle of the starting cone	/grade/:		20.0000	
Direction of worm lead	:		j	
Designed tool data				
Modules	/mm/:		4.5000	
N° of starts	/mm/:		1.0000	
Diameter of the reference cylinder	/mm/:		74.0500	
Max. diameter	/mm/		84.8500	
Axial pitch	/mm/		14.1372	
Lead	/mm/:		14.1372	
Lead angle on the tool	/grade/		3.4776	3°28`39``
Difference of lead angles	/grade/		0.0693	0°4`9``
Total tooth height	/mm/		10.8000	
Value of backing-off	/mm/		3.7463	
Lead of chip-groove	/mm/		3828.1340	
Length of the starting cone	/mm/		26.7055	
Total length of the tool	/mm/		70.5154	
Profile angle on the left side	/grade/		19.9346	19°56`4``
Profile angle on the right side	/grade/		20.0658	20°3`56``
Max. size of grinding wheel				
The required value of relief				. 16 [°]
Allowable max. diameter of the grinding wheel				89 [mm]
Correction data for resharpening of generator milling cutter for worm gear				
N° of starts on worm		[mm]		1
Modules		[mm]		4.500
Diameter of the worm reference cylinder		[mm]		72.600
Diameter of the tool reference cylinder		[mm]		74.050
Addendum cylinder diameter of the tool		[mm]		84.850
Backing angle of the tool		[°]		8.000
Value of the step in direction of the diameter		[mm]		0.100

Figure 4.38a *Design of Archimedian type worm gear generator milling cutter; input data*

Sizes of tooth in normal section		
Radius [mm](R_{gz})	Face Width [mm](S_{gz})	
31.625	10.002	
32.125	9.639	
32.625	9.276	
33.125	8.913	
33.625	8.549	
34.125	8.186	
34.625	7.823	
35.125	7.459	
35.625	7.096	
36.125	6.733	
36.625	6.369	
37.125	6.006	
37.625	5.643	
38.125	5.280	
38.625	4.916	
39.125	4.553	
39.625	4.190	
40.125	3.826	
40.625	3.463	
41.125	3.100	
41.625	2.736	
42.125	2.373	

Number	Addendum cylinder diameter (d_{a1})[mm]	Reference cylinder diameter (d_{01}) [mm]	Changing in centre distance (Δa)[mm]	Value of angle correction($\Delta\gamma$) [grade-minute-second]		
1	84.850	74.050	0.725	0 grad	4'	9"
2	84.750	73.950	0.675	0 grad	3'	53"
3	84.650	73.850	0.625	0 grad	3'	36"
4	84.550	73.750	0.575	0 grad	3'	19"
5	84.450	73.650	0.525	0 grad	3'	2"
6	84.350	73.550	0.475	0 grad	2'	45"
7	84.250	73.450	0.425	0 grad	2'	27"
8	84.150	73.350	0.375	0 grad	2'	10"
9	84.050	73.250	0.325	0 grad	1'	53"
10	83.950	73.150	0.275	0 grad	1'	36"
11	83.850	73.050	0.225	0 grad	1'	18"
12	83.750	72.950	0.175	0 grad	1'	1"
13	83.650	72.850	0.125	0 grad	0'	44"
14	83.550	72.750	0.075	0 grad	0'	26"
15	83.450	72.650	0.025	0 grad	0'	9"
16	83.350	72.550	-0.025	0 grad	0'	9"
17	83.250	72.450	-0.075	0 grad	0'	26"
18	83.150	72.350	-0.125	0 grad	0'	44"
19	83.050	72.250	-0.175	0 grad	1'	2"
20	82.950	72.150	-0.225	0 grad	1'	19"

Figure 4.38b *Calculation of the milling cutter according to block diagram; output data*

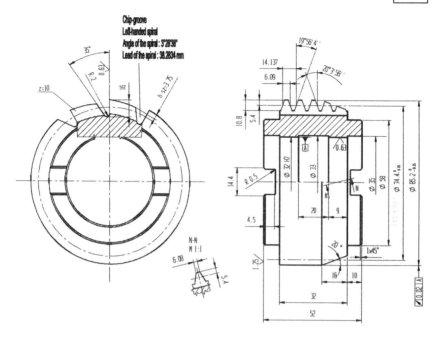

Figure 4.39 *Generator worm milling cutter*

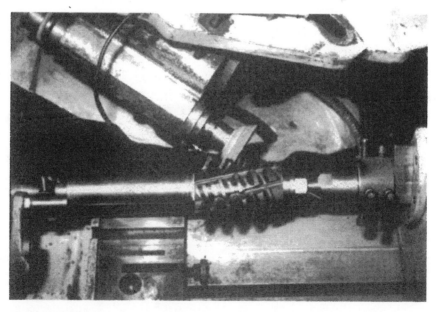

Figure 4.40 *Grinding of the generator worm milling cutter for worm having circular profile in axial section*
$m = 6.25$ mm; $K = 40$ mm; $\rho_{ax} = 25$ mm

Figure 4.41 *Grinding of the generator milling cutter for worm having circular profile in axial section*
$m = 12.5\ mm;\ K = 59\ mm;\ \rho_{ax} = 50\ mm$

Figure 4.42a *Checking after resharpening of the generator milling cutter*

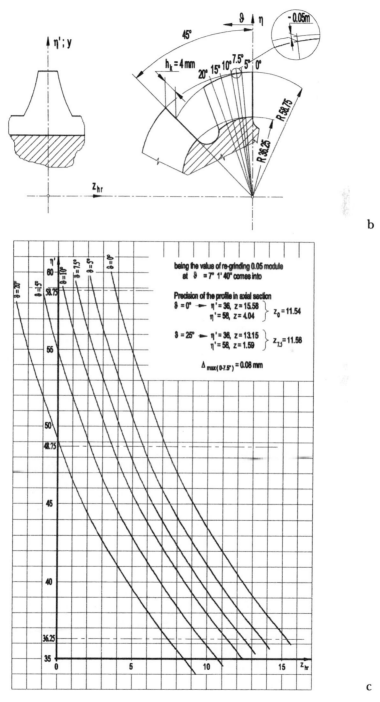

Figure 4.42b, c *Profiles calculated after the resharpening*

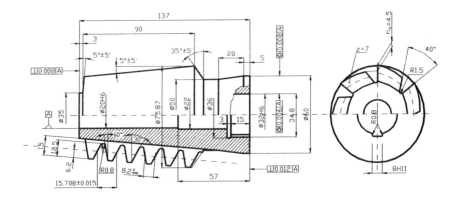

Figure 4.43a *Geometry of the generator milling cutter for spiroid gear*

Figure 4.43b *Relief by grinding of the generator milling cutter for spiroid gear*

a

b

c

Figure 4.44a, b, c *Group of manufactured tools and driving elements*

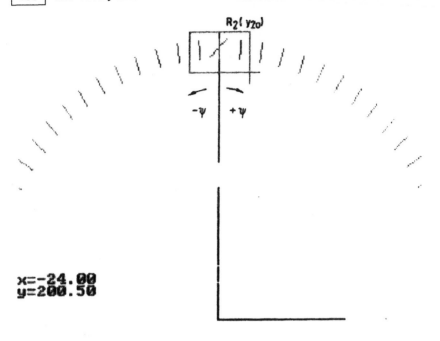

$x=-24.00$
$y=200.50$

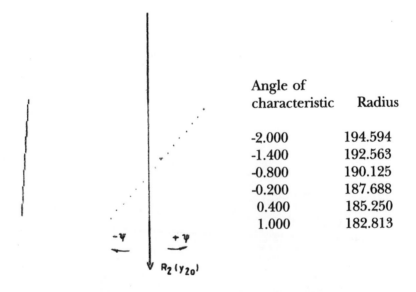

Angle of characteristic	Radius
-2.000	194.594
-1.400	192.563
-0.800	190.125
-0.200	187.688
0.400	185.250
1.000	182.813

Figure 4.45 *Contact curve and its enlarged detail with the calculated values in polar coordinates*
(The contact curve presents the points of contact for grinding of the ZK2 type worm (according to Figure 4.4e))

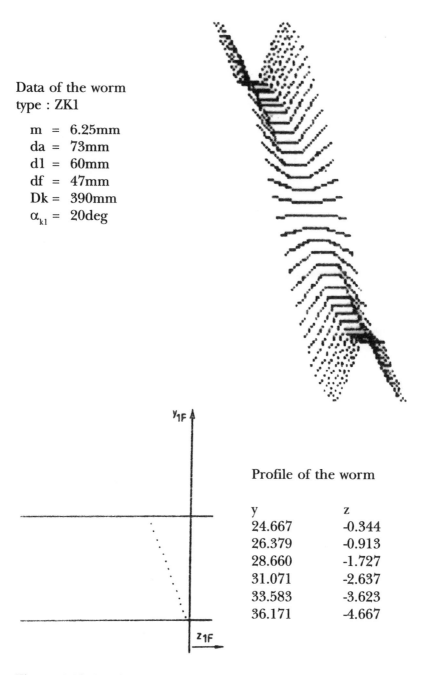

Data of the worm
type : ZK1

m = 6.25mm
da = 73mm
d1 = 60mm
df = 47mm
Dk = 390mm
α_{kl} = 20deg

Profile of the worm

y	z
24.667	-0.344
26.379	-0.913
28.660	-1.727
31.071	-2.637
33.583	-3.623
36.171	-4.667

Figure 4.46 *Coordinates of the profile points in the axial section of the ZK2 type worm (according to Figure 4.4e)*

5

GRINDING WHEEL
PROFILING DEVICES

The exact mathematical investigations of geometric problems can only achieve their aims if they can be validated in practice in the manufacturing process.

During the past decades several different grinding wheel profiling devices (regulators) for manufacturing helicoidal surfaces were developed. We shall now summarize the features of these in order that they may be differentiated according to the principle of regulation, either as mechanical generators equipped by rigid programming or as numerically controlled CNC equipment.

Though both types of devices serve the same purpose, they operate basically on different principles. These two principles are:

1. Principle of backward generation: The tool imitating the theoretical surface of the worm with the vertex point of a diamond regulates the grinding wheel profile. This is the working principle of traditionally used additional grinding wheel profile generating devices (as in ruled surface worms) which, positioned in place of the worm and joined into the kinematical chain of the machine tool, imitate the geometric shape of real worm.

   Assuming that diamond is used as the engineering material for the helicoidal surface, the grinding wheel is ground to a correct profile within the kinematical chain, in the actual position of the mated worm.

2. Principle of enveloping: This principle differs from the principle of backward generation in that the grinding wheel profile to be generated or the worm profile to be developed is determined by calculations assuming that the tool and worm mutually envelop each other.

This principle can be applied far more widely than the previous one. Suitable software can be applied for arbitrarily-mated tool helicoid surfaces. The principle of enveloping involves theoretical kinematical considerations while the principle of backward generation requires more practical treatment.

5.1 DEVICES OPERATED ACCORDING TO MECHANICAL PRINCIPLE

The most used mechanically operated or rigid program holder devices can be classified into the following groups:

1. The control program is embodied in the thread-copying gauge (eg device regulation in axial section of the grinding wheel developed by the firm Klingelnberg).
2. Copying by pantograph (eg the 'Diaform' equipment) (Klingelnberg).
3. Devices realizing 'control by generation' in the axial section of helicoidal surfaces applicable to ruled surfaces ('matrix' procedure) (Csibi, 1990).
4. Devices to regulate the wheel along its arc on grinding wheels for arched profile worms (Niemann, Litvin method) (Patentschrift, Deutsches Patentamt, Nos 905444 47h 3/855527 27h; Russian Patent No. 139.531).
5. Device to regulate the wheel in preparation of its profile for precise grinding in the axial sections of the circular profile worm. This device has been developed and used by the author (Dudás, 1973; Dudás et al, 1983).

Of the solutions enumerated above, the author's own research work (5) will be described. The structure of this device and principle of its operation will be summarized briefly. The other important shape-correcting grinding processes have already been reviewed in Chapter 2. There it was noted that the problems of worms with circular profiles are related to problems of worms having ruled surfaces. This is important because the shape of a grinding wheel, and whether it is precise enough, determines the distortion in the worm, especially for high leads. This fact is emphasized in the literature (Niemann and Weber, 1954).

Therefore, determination of the profile of the wheel so that no distortion of the worm surface could occur either in axial or in normal section, and thus keep the required (the prescribed) profile, was the aim of the work.

The essence of the operational principle of the device designed by the author is that the radius ρ_{ax} of the generator circle of the worm to be ground is moved into a well-defined position, making it possible for it to roll down before the grinding wheel (Dudás, 1986a). The grinding wheel should be turned away by angle γ (generally the mean lead angle on the workpiece to get it in working position) while the regulating device should be positioned into the centre line of the main spindle (this is the centre line of worm) (Dudás, 1973; Dudás, Csermely and Varga, 1994).

The regulating device (Figure 5.2) is centred (III-III axis) through its driving fork joined to the main spindle, and the kinematic chain Q of the machine tool manages its axial displacement in accordance with the parameter p and it turns parallel to the grinding wheel as well as the worm to be machined.

The wheel is regulated by turning it grade by grade up to completion; the diamond, together with the generating regulator, roll out to leave the wheel surface. In Figure 5.1 diamond No 1 shapes up the addendum circle of the wheel while diamond No 2 manufactures the working surface of it (ρ_{ax}).

Figure 5.3 shows the process of wheel generation in practice. The diamond No 2 vertex point (Figures 5.1 and 5.3) is joined to the guiding arm so that its centre of rotation is distance K from the worm axis. The profile of the wheel obtained by generating regulation was compared with the wheel profiles obtained without regulation. The result is shown in Figure 5.4.

Figure 5.4 shows that by generation, the details (addendum) of the wheel that cause the profile distortion of the worm in normal section during machining carried out during roll in and out are removed.

Profile distortion caused by change of the wheel size (Hoschek, 1965; Niemann and Weber, 1954) when applying the new generating method does not occur. That is, the helicoidal surface obtained by machining is independent of wheel diameter, because the wheel profile was obtained by backward generation from the worm in every case of re-regulation (Dudás, 1980; Dudás and Ankli, 1978).

This principle is important because the tool required to generate the worm gear can be manufactured corresponding to the worm with which it will be mated.

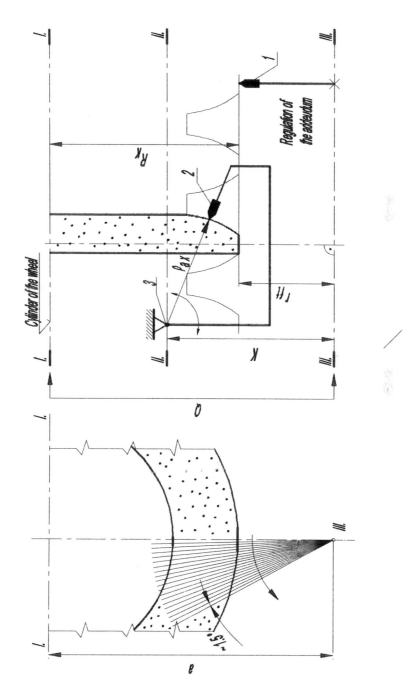

Figure 5.1 *Principle of operation of the generating regulator device (Dudás type)*

5.2 ADVANCED VERSION OF THE WHEEL-REGULATING DEVICE OPERATING ON THE MECHANICAL PRINCIPLE

For mass-production manufacturing it is necessary to develop a wheel-regulating device suitable for universal use over a wide range of sizes. In previous versions, separate bridles were necessary for the different sizes with parameters being changed; this implies additional manufacturing capacity. To avoid these difficulties the device shown in Figure 5.5 was developed. Its operation is described below (Dudás and Ankli, 1978).

Body 14 is linked to the centres so that the device is connected into the Q kinematical chain. Element 12 is to be shifted (after loosening bolts 17) to adjust the centre of the circular arc guiding the diamond point through distance K. The parameter ρ_{ax} can be adjusted by choosing the centre of rotation for the teethed stirrup frame 8 and by shifting axially the diamond 2. For smaller sizes of ρ_{ax} it is better to use the point 18 on body 12, for the centre of rotation to apply a suitable size stirrup frame. Mating the involute worm having two starts with the helical gear produces the generating movement. The spindle 6 is mated with gear 7, which is mated to the helical teethed stirrup frame 8.

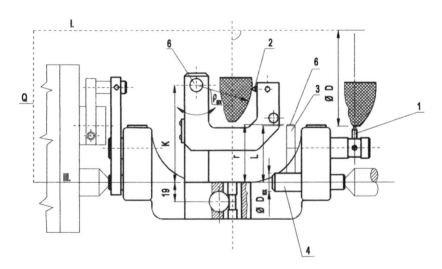

Figure 5.2 *Sketch of operation of the regulated generator device (Dudás, 1976)*

The adjustment limits are K = 50–100 mm and ρ_{ax} = 20–60 mm.

This device has been used in a significant number of worm gear drives produced recently. Practical experience has testified to its ease of use, its excellent precision and the improvement in quality in surface roughness of worm acheived.

The diamond point (2) runs through several generator lines situated side by side on the thread side of the worm and regulates the grinding wheel profile correctly from a kinematic point of view. The diamond point (2) is joined to the grinding arm so that its centre of rotation is distance K from the axis of the worm. The value of the circular profile (ρ_{ax}) in the axial section of worm can be precisely adjusted in the device. The grinding wheel regulated by this method for the grinding of the worm generates a kinematically correct worm profile.

The profile of the wheel obtained by generating regulation has been compared with the profiles of wheels obtained without use of generating regulation. The difference between them can be seen in Figure 5.6.

A finer finishing wheel can be generated by shifting the eccentric arm into a 1:2 gearing ratio position and the movements can be produced by position 9, 6, 7 and 8 connections. Otherwise, the operation of this device is the equivalent of that previously described (see Figure 5.2).

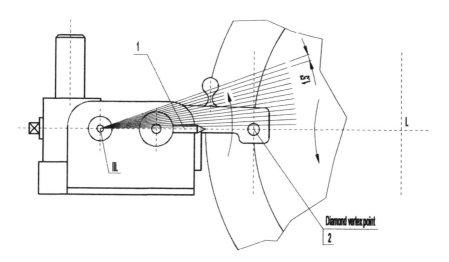

Figure 5.3 *Principle of regulation by wheel generation (Dudás, 1973)*

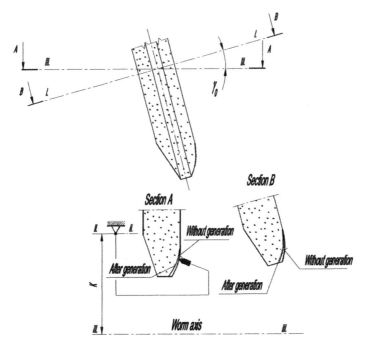

Figure 5.4 *Wheel profiles obtained by regulated generator and without regulation*

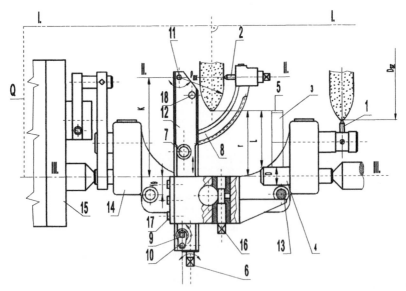

Figure 5.5 *Operation of the advanced version of the wheel-generating device (Dudás and Ankli, 1978)*

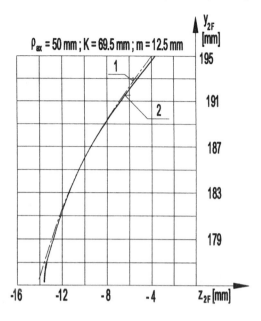

Figure 5.6 *Actual wheel profiles obtained using generating regulation (2) and without generating regulation (1) (a = 200 mm, z_1 = 3, m = 12.5 mm)*

Figure 5.7 *Generation of the wheel at the end of generation in the rolling out position of the profiling device (test manufacturing) (Dudás, 1973, Dudás et al, 1983)*

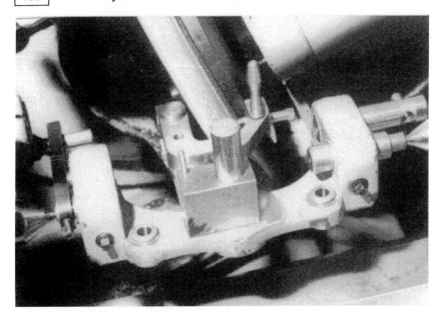

Figure 5.8 *Wheel profiling device in working position (machining of ρ_{ax})*

Figure 5.9 *The advanced version of the wheel profiling device at commencement of generation (Dudás and Ankli, 1978)*

Figure 5.10 _Practical realization of grinding_

5.3 CNC-CONTROLLED GRINDING WHEEL PROFILING EQUIPMENT FOR GENERAL USE

In this section the equipment and the procedure that make possible the profiling of grinding wheels suitable for machining geometrically correct planes and axisymmetrical and helicoidal surfaces are discussed.

For worm gear drives transferring high power and of high gearing ratio, it is necessary to shape up the contact surfaces to facilitate high load-carrying capacity and allow the build-up of hydrodynamic conditions of lubrication. These contact surfaces are characterized by the fact that the majority of momentary contact curves are inclined by at about 90° to the relative velocities within the field of engagement (Niemann and Weber, 1942).

Besides the better conditions of engagement brought about by developing geometrically correct grinding technology it should be possible that changes in diameter of the grinding wheel, after each re-regulation, do not influence the shape of the worm but allow it to always generate the same helicoidal surface. It is important that the grinding of the tool, which will be used for generation of the worm gear, can also used to prepare the equivalent worm with which it will be mated.

To fulfil the above requirements so that:

- removal of workpiece from the machine, necessary for regulation of the wheel, and then clamping it again, is avoided;
- it is possible that the grinding of the wheel with changing profile be regulated with at least the precision determined by the precision limit (during operation);
- the process is able to fit or to integrate into a highly automated manufacturing system;

it is best to use a device controlled by CNC.

The aim of the CNC device is to avoid the mistakes and shortfalls of traditional procedures and machine tools, to reach better engagement conditions than at present and to create worm gear drives of high efficiency. Such a device, owing to its numerical control being able to be integrated into computer aided manufacturing systems, operates with high precision and easily compensates for profile distortions automatically. It is equally suitable both for time-to-time and for continuous profiling. The profiling device, being independent equipment, can be linked to existing traditional or to numerically-controlled grinding machine tools or it can be built into the unit. To realize these aims, the principle of enveloping instead of the principle of backward generation should be used. Both principles were discussed at the beginning of this chapter.

The grinding wheel, regulated by this equipment, can be regarded as backward generated from the worm to be ground. Because the profile of the grinding wheel is determined as an enveloping of that to be machined using the mathematical method, the wheel will be profiled after that. Both the mathematical and the numerical algorithms developed make possible the determination of tool profiles suitable for machining arbitrary cylindrical or conical helicoidal surfaces.

The calculations should be repeated or suitable corrections made when the centre distance of the grinding wheel and machined surface has changed.

The wheel-regulating device can be built as one of several constructional variants. For one of these constructions (see Figure 5.15) the grinding wheel regulating device operates by angular displacement of the diamond tool (pencil) round axis B1 (perpendicular to the surface) and will dress the wheel controlled along two axes

Figures 5.11 – 5.12 *Wheel regulation (suitable for grinding of spiroid worm) of cubic boron-nitride wheel (E63 150x10x32x12x2, LO 63/50 C2 K)*

to produce the required shape. This solution requires NC control along three axes (Su and Dudás, 1966).

Another possible grinding wheel regulation device (Figure 5.16) is shaped up so that the diamond disc, of machining grinding wheel profile, rotates round its own axis while it is controlled, according to the required profile, along the two axes $(x_1 - z_1)$.

This type, the regulating roller device, is suitable for controlling the 2D path parallel to the correction of the tool radius.

It is suggested that by turning the wheel round axis B2, in accordance with Figures 5.16, 5.17 and 5.18, the field of the application of this device would be widened.

In Figure 5.17 the two-roller regulating devices can be seen. In the device shown in Figures, 5.16 and 5.17, the CNC axes (suffixes 2) give the directions of relative displacement between the grinding wheel and the CNC-controlled grinding machine tool.

It was planned to develop a device meeting several aims (Figure 5.18). The advantage of this solution is that both sides of the grinding wheel can be profiled by using the same tool. One possible variant is a construction using a single wheel.

It provides the possibility of formation using a simpler construction. The control system is similar to that applied in conventional NC lathe turning machines.

It differs basically from all existing devices in that, when adjusting it to an optimal angle (α_{opt}), one side of the wheel is kept constant during machining, both sides of the grinding wheel being continuously regulated with the same roller and it being turned round axis B_2. (The value of the turning angle during machining will be optimized by calculation.) Applying this method, it is possible to manufacture a workpiece with arbitrary profile with minimum wear of the wheel.

By continuous turning it round axis B_2 (round axis x_2), it can be guaranteed that, by minimal re-regulation of the wheel, owing to its continuous regulation, the result will be a geometrically correct profile. The maximum value of turn, (B_2) max, is the value of angle corresponding to difference Δz_2 (see Figure 3.33).

This device can be applied in two versions, partly as an additional automation unit for existing machines (Figure 5.19), and partly as a unit of a newly constructed thread grinding CNC machine (Figure 5.20). Continuous regulation is only possible in the second situation.

The symbols, applied in previous figures, are:

SZB	–	regulating device
TE	–	power electronics unit
NC	–	numerical controlling unit
ka	–	data supply manual or it is solved by auxiliary data bank
vj	–	controlling signal
ej	–	reference signal
INT	–	interface unit
PC	–	personal computer
MO	–	screen unit
TU	–	keyboard unit
HT	–	(magnetic) auxiliary data bank
PR	–	printer unit
NC/PC	–	numerical control or personal computer
NCK	–	numerical control unit of the grinding machine
TEK	–	power electronics unit to grinding machine
KG	–	grinding machine

The block diagram related to generation of CNC control statements and profiles of grinding wheel can be seen in Figure 3.32.

Figure 5.13 _Grinding of KI type spiroid worm using grinding wheel having superhard grains_
_($m = 5$, $z_1 = 1$, $a = 100$ mm, $r_D = 6$ mm (Dudás, 1986d)_

Figure 5.14 *Grinding of KI type spiroid worm using grinding wheel having superhard grains*
(E63 150x10x32x12x2 LO 63/50 C2 K)

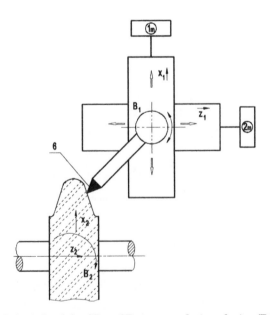

Figure 5.15 *Principle of the 2L + 1R type regulating device (Dudás, 1988a)*

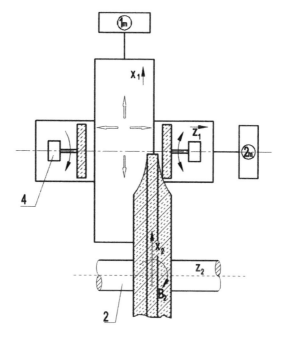

Figure 5.16 _Principle of the 2D (2L) type regulating device (Dudás, 1988a)_

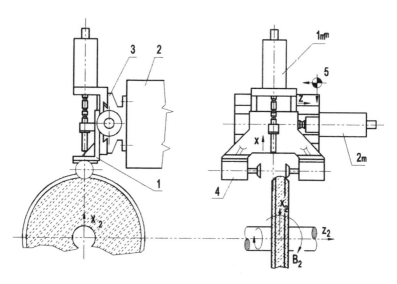

Figure 5.17 _One possible variant of two-roller 2D regulating device (Dudás, 1988a)_

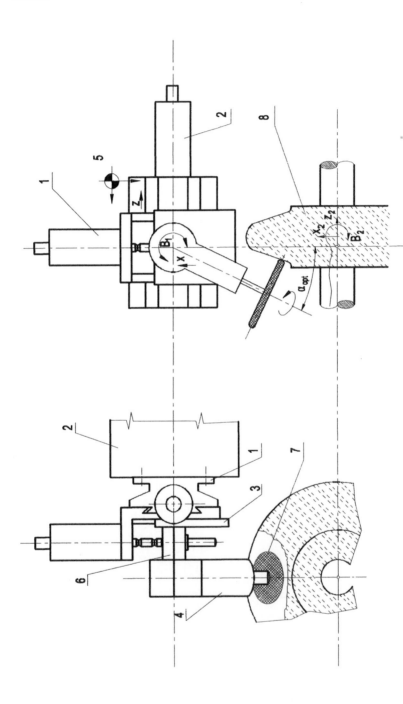

Figure 5.18 *The new wheel regulating device (Dudás, 1988a)*

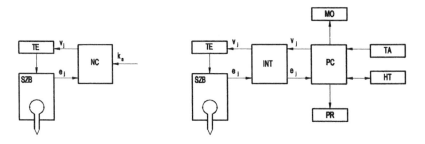

Figure 5.19 *Block schemes of the two possible variants to apply periodical wheel-regulating device*

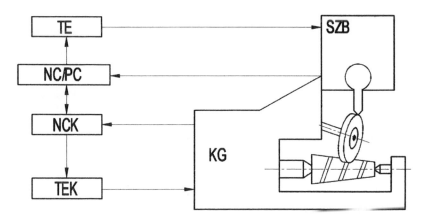

Figure 5.20 *Block scheme of the continuous wheel-regulating device (Dudás, 1988a)*

6

QUALITY CONTROL OF WORMS

The assessment of worm gear drives is a many-sided task, including the checking of the geometric sizes of the mated elements, the analysis of their meshing conditions and investigation of their operational characteristics. Expert evaluation of a worm gear drive is only possible after completing a series of tasks and evaluating their results.

Assessment should cover the following areas:

- checking geometry of the worm;
- investigation of the meshing conditions (see Chapter 8);
- checking the more important characteristics of the operation (see Chapter 8).

6.1 CHECKING THE GEOMETRY OF WORMS

Geometric checking should include:

- deviation in axial pitch f_{pl};
- deviation in cumulative lead f_{pxr};
- deviations of worm profile f_{fr};
- radial runout of the worm thread line F_{rlr}.

The teeth of the worm can be checked using a special checking device made for the checking of hobs, by using a universal device for the complex checking of worm gear drives or by using universal measuring instruments or devices providing measurement of worm thread lines.

The advantage of the special device for checking modular hobs is to make quick and simple checking possible. Its disadvantage is that it can only be used for worm checking. To check the worm and the worm gear separately as well as to check the drive during the operation can only be carried out using universal measuring devices.

The checking of the above-mentioned parameters of the worm gear drive was carried out using a PSR-750 type (Klingelnberg made) complex worm gear drive checking device. All the measurements carried out are not reviewed here but in the section dealing with measurements of profile deviations of the worm. Because it is closely connected with manufacture and inspection of geometrically coupled worm–worm gear it may be carried out as suggested in Chapter 2.

The main characteristics of the worms manufactured by the author are summarized in the table shown in Figure 4.6.

6.1.1 Determination of worm profile deviation

The deviation of a worm profile is the perpendicular distance (f_{fr}) between two theoretically correct tooth profiles spanning upper and lower extremities of the actual profile of the worm within the working range (Figure 6.1). The precise solution in the case of a circular profile can be found in (6.1).

The theoretical bases of this investigation are presented in Figures 6.1 and 6.2. Measurement of profile deviation should be carried out in the axial section of the worm, ie the plane of the nominal circular profile. The deviations of the theoretical and the actual profiles from line a–a should be determined; the differences of the value of profile deviation can then be determined. The line a–a is a chord perpendicular to normal at the point of the theoretical profile fitted on the reference cylinder.

The difference between theoretical and actual profiles is:

$$\Delta_h = h' - h \tag{6.1}$$

where:

h' is arc depth on actual profile (as determined by measurement),
h is arc depth on theoretical profile (as determined by calculation).

The value of profile deviation is determined by the sum of the absolute values of the upper limit deviation (plus) and lower limit deviation (minus), that is:

$$f_{fr} = \left|\Delta^+_{hmax}\right| + \left|\Delta^-_{hmin}\right| \qquad (6.2)$$

The actual value of this deviation should be less than the tolerance of profile deviation f_f (MSZ 05.5502–75).

The distance h_i between the theoretical profile and line a–a can be calculated using Figure 6.2.

The equation giving the correlation between arc height h_i and chord length S_i based on Figure 6.2. is:

$$h_i = \rho_{ax} - \frac{1}{2}\sqrt{4\rho_{ax}^2 - S_i^2} \qquad (6.3)$$

where i = 0, 1, 2, 3 ...

Using equation (6.3) to obtain the arbitrary a–a line, the theoretical values of $\pm\frac{S_i}{2}$ and the arc height h_i can be calculated.

After completing the calculations, values of h' can be determined by measurement, guiding the preceptor along line a–a. Then the value of the worm profile deviation can be obtained. The theoretical layout of the measurement is given in Figure 6.3.

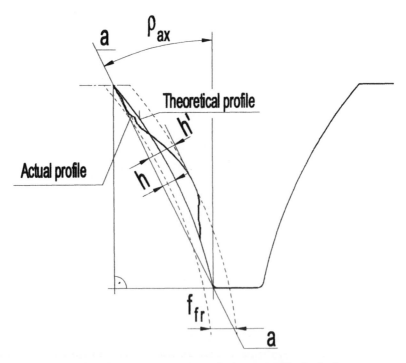

Figure 6.1 *Interpretation of worm profile deviation*

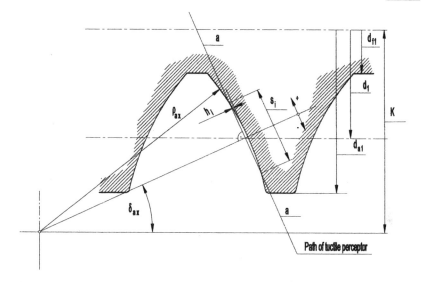

Figure 6.2 *Mathematical bases of worm profile deviation measurements*

During measurement (following the steps in Figure 6.3) the measuring head 8 has to be rotated round point O_2 by angle δ_{ax} with the help of adjusting pin 9 and by applying a set of calibrated measuring plates 5. The angle δ_{ax} refers to the perpendicular to the worm axis, at position $\delta_{ax} = 0$.

Then the preceptor 9 should be set on the reference cylinder line and the indicator 4 should be set to zero for indication displacement $\pm \frac{S_i}{2}$ and, at the same time, indicator 3 should be set to zero as well. The accuracy of these indicators is 1/1000 mm.

The feeler of the original system of the device (PSR-750) cannot be used because of chord height ($\delta_{ax1} = 50$ mm and $\delta_{ax2} = 58.5$ mm). A new feeler was manufactured to measure the profile deviation for use within the total range of the tooth flank.

The feeler deflection h_i which accords to actual profile can be measured as a function of the axial displacement $\pm \frac{S_i}{2}$ by adjusting screw 6.

In the example tabulated in Figure 6.4 are shown the results of the measurement carried out on a worm with gearing ratio $i = 11.67$. The table contains both theoretical and measured values, their differences and the allowable deviations within the tolerance (f_r).

The maximum profile deviation calculated from the values shown in the table is:

Measured parameter and position	Calculated values h [mm]	Measured values H' [mm]	Difference Δ_h [mm]	Tolerance f_f [mm]
h_1	0.0405	0.033	+0.075	
$-\dfrac{S_1}{2}$	2	–	–	0.08
h_2	0.1605	0.146	+0.0145	
$-\dfrac{S_2}{2}$	4	–	–	According
h_3	0.3315	0.341	-0.0095	to standard
h_1	0.0405	0.040	+0.0005	
$+\dfrac{S_1}{2}$	2	–	–	TGL 17266
h_2	0.1605	0.148	+0.0125	and
$+\dfrac{S_2}{2}$	4	–	–	
h_3	0.3315	0.318	+0.0135	MSZ 05.5502-75
$\pm\dfrac{S_3}{2}$	6	–	–	
Direction of ± according to Fig. 6.2.				

Figure 6.3 *Sketch of the profile deviation measurement*

$$f_{fr} = \left|\Delta_{hmax}^{+}\right| + \left|\Delta_{hmin}^{-}\right| = |0.0145| + |-0.0095| = 0.024 \text{ mm}$$

Based on the tabulated values (Figure 6.4) it can be seen that the worm was manufactured well within the tolerance allowed (0.08 mm).

The results confirm that, with the circular generator curve of the worm tooth positioned in its axial section, together with use of the generating method to shape up profile of the disc grinding wheel and the generator hob, as was discussed in Chapter 2, the method proved to be satisfactory.

6.2 CHECKING OF HELICOIDAL SURFACES ON 3D MEASURING MACHINES

Distortion and shape deviations of helicoid surfaces during machining can occur owing to wear and re-sharpening of the tool employed, together with limited accuracy in the setting of the tool.

The geometrical checking methods used for helicoidal surfaces have changed during the years; they have been updated and are now becoming more accurate. Master gauges previously used to check profiles, the go and no go gauges, made possible the assessment of surfaces on a subjective basis and could only be used for helicoid surfaces with well-defined geometric characteristics. Up-to-date devices, employing

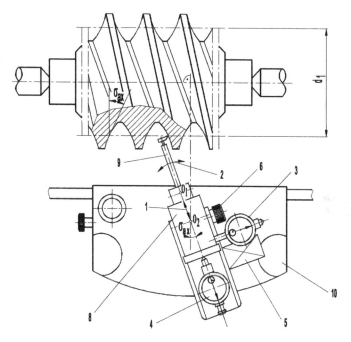

Figure 6.4 *Profile deviations as defined in Figure 6.2*

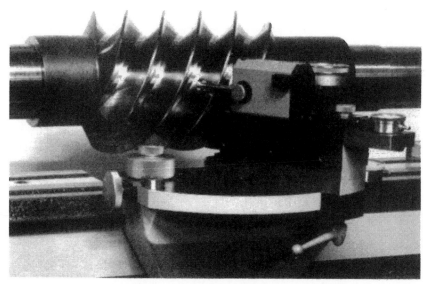

Figure 6.5: *Measurement of worm profile shape deviation using Klingelnberg PSR-750 type measuring device*

high-precision measurement gauges, can now provide assessment on a more objective basis, with 0.01 – 0.001 mm accuracy.

The basic problem of the traditional geometrical checking methods is to handle helicoidal surfaces as two-dimensional plane figures and to evaluate the cumulating effects of different directional deviations as planar effects (eg axial pitch, effect of the radial runout in face plane, profile deviation in axial section etc).

The problem is particularly outstanding in the evaluation of the profile deviations of helicoidal surfaces.

> ISO 1328–1975 defines profile deviation as: 'The perpendicular distance between two neighbouring nominal profiles limiting the actual profile on two sides at a given section
> along the working profile'.

The limits of this definition in the case of the circular profile worm (between two arcs erected from different centre points, to define the perpendicular distance) are especially problematical (Figure 6.1).

Therefore it is essential that there is a theoretically laid basis for the checking method for worm surfaces as three-dimensional blocks, their detailed working out and use in practice.

To answer this demand, the application of computer techniques to measurement has made it possible to handle a database efficiently by connecting coordinate measuring devices to computers. These three coordinate measuring devices have already been applied to several levels of automation.

Efforts have been made within recent years to carry out the geometric checking of helicoidal surfaces with high precision using three coordinate measuring devices. To make them suitable to this task they are equipped with NC- or CNC-controlled high-precision ancillary equipment (eg indexing attachments, mechanized measuring heads, etc) and with respect to their construction can be regarded as single-purpose equipment. For this reason their use can only be regarded as economic for mass production. Leading manufacturers strive to integrate the three-dimensional measuring systems into manufacturing the line.

6.2.1 Use of 3D measuring machines

The system of conditions for the operation of a measuring system is determined basically by the characteristics of the machine in question. The important characteristics are presented in Figure 6.6.

- ■ Manual equipment: hand-operated and controlled;
- ■ motor-driven equipment: hand-controlled with servo-driven movement transmitters. Protected against overload of stylus;
- ■ CNC equipment: controlled by microprocessors so the complete process of measurement can be automated.

The level of automation of the measuring equipment determines to a great extent the methods of checking for different surfaces, and the design of the software handling the data measured.

6.2.1.1 CNC-controlled equipment supplied with circular table

An additional unit situated on the measuring table, in practice a circular table, guarantees high-precision positioning. The master-piece is fixed on the circular table and by touching it with the surface of the stylus, this master-piece, using a 'teaching program', generates in the memory of the computer the path performed by the feeler point, which is a surface of equal distance to the surface

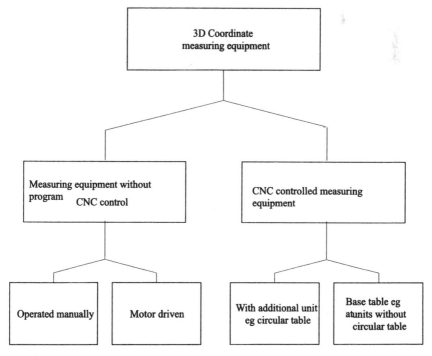

Figure 6.6 *Classification of measuring machines according to level of automation*

of the master-piece. Then the workpiece to be checked has to be placed in the same position, so that the measuring head can be directed into known points and their deviation from measured points can be determined. Using statistics to evaluate the deviations, the helicoidal surface can be assessed (using a large number of points). In the case of small-scale manufacturing it is not worth manufacturing a master-piece; instead, a separate program can be used to determine coordinates of the points to be checked, in the form of the vector control facing in the direction of the surface normal.

6.2.1.2 CNC-controlled equipment supplied with non-circular basic table

This method does not require the fixing of the workpiece to be checked in a defined position on the measuring table. The 'scanning' technique is most frequently used to define the surface to be measured. The essence of this method is that the measuring head is controlled in two directions of the coordinate system, while in the third direction the automated stylus follows the surface. The surface determined by the centre point of the spherical end of the stylus in well-defined pitches will be determined, so any arbitrary surface can be surveyed by the system. To determine the actual surface from the surface described by the centre point of the spherical stylus is only possible approximately. An approximate determination of the actual surface points can be carried out as follows:

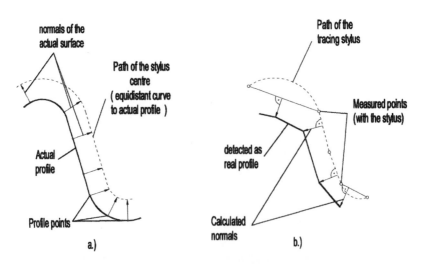

Figure 6.7 *Determination of real tangential points*

1. First, the equalizing curves or an equalizing surface will be fitted to the set of points determined by the spherical stylus (see Kyusojin *et al.* 1986).
2. Second, along the normal vectors of the equalizing surface the points of the actual surface can be defined as points situated at distance r, the radius of the spherical stylus used.

The approximate nature of this procedure comes partly from the approximate character of the equalizing surface, and partly from the fact that the point on the actual surface is determined on the normal vector of the previous, equalizing surface (Figure 6.7b), though the centre point of the spherical stylus is situated along the normal vector of the real surface (Figure 6.7a).

6.3 CHECKING OF HELICOIDAL SURFACES BY APPLICATION OF 3D MEASURING DEVICE PREPARED FOR GENERAL USE (WITHOUT CIRCULAR TABLE, CNC-CONTROLLED)

As with helicoidal surfaces, these are always manufactured, except in specialist manufacturing firms, in small or medium size quantities, the capacity required for checking them being less than the capacity of the measuring machines. The automated measurement of helicoidal surfaces, taking into account new developments, could be used with, for example, a commercial 3D measuring device.

Following is a description of a method of potential general use for the checking of the geometry of helicoidal surfaces. This takes into account the characteristics of coordinate measuring devices (eg Mitutoyo, DEA, Johanson, etc types) (Bányai, 1977).

The own-program system of the general measuring device is suitable for coordinate transformations, so the helicoidal surface can be measured in its own coordinate system and the checking by measurement can be carried out in that coordinate system. Using a ruby sphere as a stylus, the counter of the measuring device, at the point of contact of the stylus and the surface, will register the coordinates x, y, z of the spherical stylus centre (Figure 6.8). The coordinates of the contact point would be determined by calculations that influence the precision of the measurement.

In developing the measurement program it should be an aim that the geometric checking is carried out at uniform points independent of the type of helicoidal surface. To fulfil this requirement the equations of the different types of helicoidal surfaces ($\vec{r}_{1F}$) should be formulated in common bases as in Chapter 3.

By appropriate selection of the parameters of helicoidal surfaces the possibility arises to detect the origin of errors shown up in the geometric checking. From a practical point of view it is useful to choose as a parameter for the helicoidal surface one of the coordinates determining the generator curve (edge of the tool, contact curve etc) of the helicoid surface (η).

It can therefore be concluded, and as seen in Chapter 3, the equations for different types of helicoid surfaces can be determined from the data. In the general case it can be written as:

$$\vec{r}_{1F} = \vec{r}_{1F}(\eta, \vartheta) \tag{6.4}$$

It is known that the normal vector at the contact point of a sphere and any oblique surface should fit on to the centre point of the sphere. This can be used to determine the contact point of the theoretical and actual helicoidal surfaces as well as their inclination (into the direction of the normal vector). To solve this problem the normal vector of the theoretical helicoidal surface should be known in accordance with equation (3.73). Utilizing the fact that the line of the normal vector should fit on the centre point of the spherical stylus $\vec{r}_t(x, y, z)$ the contact point of the theoretical helicoid surface can be determined.

This point is the piercing point of the line on the above-mentioned normal vector $\vec{n}$ and the theoretical helicoidal surface, and would be given by simultaneous solution of the two equations:

$$\left. \begin{array}{c} \dfrac{\vec{n}}{|\vec{n}|} = \dfrac{\vec{r}_t - \vec{r}_{1F}}{|\vec{r}_t - \vec{r}_{1F}|} \\[2em] \vec{r}_{1F} = \vec{r}_{1F}(\eta, \vartheta) \end{array} \right\} \tag{6.5}$$

This system of equations, taking into consideration the deductions drawn in previous chapters, is a transcendental function with respect to parameter ϑ but it can be solved by applying a numerical algorithm. The coordinates of the contact points of the theoretical helicoidal surface (x_{1F}, y_{1F}, z_{1F}) can be obtained by substituting the parameters η and ϑ, tied to each other, into the equation of the investigated helicoidal surface.

The real contact point can be determined as shown in Figure 6.8 and sits on the sphere of the stylus so it is at distance r from the centre point of that sphere. The components of this normal distance, in the direction of the coordinate axes, are Δx, Δy, Δz, varying according to the position of the contact point. The coordinates of these points are:

$$\left.\begin{array}{l} x_m = x + \Delta x \\ y_m = y + \Delta y \\ z_m = z + \Delta z \end{array}\right\} \tag{6.6}$$

The deviation between the theoretical and measured points is:

$$\left.\begin{array}{l} \delta_x = x_{1F} - x_m \\ \delta_y = y_{1F} - y_m \\ \delta_z = z_{1F} - z_m \end{array}\right\} \tag{6.7}$$

It is to be expected that the result of mathematical calculations should differ from those using the actual contact points and deviations, so the actual helicoidal surface would differ from the

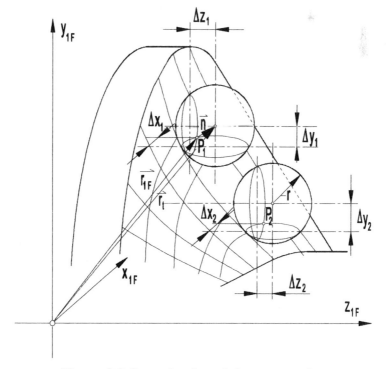

Figure 6.8 *Determination of the contact point*

theoretical one. The position of the centre point of the spherical stylus is determined by the fact that it is in the contact point position that had been calculated using the theoretical helicoidal surface. This is why the first approximation of the coordinates of the contact points was inaccurate. The calculations to determine the contact points should be repeated using the actual helicoidal surface parameters in stages, resulting in more precise evaluation of the contact points.

The most efficient way to increase the accuracy of calculated corrections is to modify the lead parameter p, because the theoretical helicoidal surface calculated using the measured value of the lead parameter gives the smallest difference compared with the actual surface. The normal vector of the theoretical helicoidal surface is determined by using its original parameters placed on point (1); P_t pierces the helicoid surface at the point P_e (Figure 6.10). The piercing point P_e', determined by the modified helicoidal surface, is far more precise than the actual contact point P_v placed on the surface of the

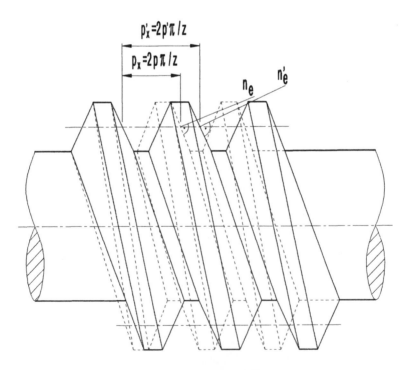

Figure 6.9 *Effect of lead error*

spherical stylus. The corrections and the program developed during the processing of measured values are carried out automatically.

The component coordinates of the distance between the contact point and the centre point of the spherical stylus are:

$$
\left.\begin{aligned}
\Delta_x &= \frac{n_x}{|n|}r = x_k \cdot r \\
\Delta_y &= \frac{n_y}{|n|}r = y_k \cdot r \\
\Delta z &= \frac{n_z}{|n|}r = z_k \cdot r
\end{aligned}\right\}
$$

(6.8)

Using these values the actual contact points can be found, together with their deviations from the theoretical points.

The correction coefficients (x_K, y_K, z_K), for the Archimedian helicoidal surface, for the direction of the normal vector ($\frac{\vec{h}}{|\vec{h}|}$) are:

$$
\left.\begin{aligned}
x_k &= \frac{p \cdot \cos\vartheta + \eta \cdot \sin\vartheta \cdot \mathrm{tg}\,\alpha_{on}}{p^2 + \eta^2(1 + \mathrm{tg}^2\alpha_{on})} = \frac{n_x}{|n|} \\
y_k &= \frac{p \cdot \sin\vartheta - \cos\vartheta \cdot \mathrm{tg}\,\alpha_{on}}{p^2 + \eta^2(1 + \mathrm{tg}^2\alpha_{on})} = \frac{n_y}{|n|} \\
z_k &= \frac{\eta}{p^2 + \eta^2(1 + \mathrm{tg}^2\alpha_{on})} = \frac{n_z}{|n|}
\end{aligned}\right\}
$$

(6.9)

The deviation due to the lead difference is shown in Figure 6.9. The lead difference on the helicoidal surface causes greater error in the calculation of the theoretical contact point as Figure 6.10 shows.

The lead difference is determined as follows: the lead, by definition, is determined on the thread curve on the reference cylinder of the helicoidal surface and the (measured) contact points do not usually fit on the reference cylinder, so they should be moved on this surface along the theoretical axial profile (Figure 6.11).

For the 'u' points transferred on the reference cylinder, (assuming) $P_e' \approx P_v$, a regression line as the function of parameter ϑ can be determined, giving the lead (p_z) characterizing the helicoidal surface as described by the points measured (Figure 6.12).

Variations in the starting theoretical value leads to lead variations. When the lead deviation exceeds the tolerance provided by

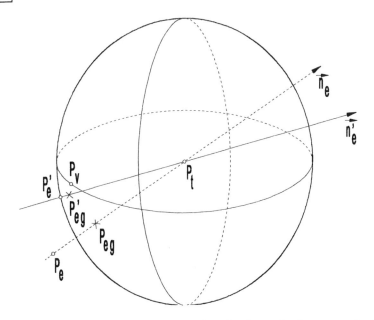

Figure 6.10 *The normal vectors are* $\bar{u}_e$, *the theoretical one used at the beginning of the calculations, and* $\bar{u}_e$, *the one corrected using the lead inclination*

precision measurement, the program repeats the calculation, taking into account the new lead parameter of the contact points.

The calculations of actual contact points can cause mistakes if the theoretical helicoidal surface is not situated where it was supposed to be, in other words, if the coordinate systems of the theoretical helicoidal surface and the chosen one do not coincide (Figure 6.13).

The elimination of the calculation error owing to the difference between the coordinate systems, the theoretical (belonging to a 'proper' helicoidal surface) and the chosen one, is easier than elimination of that caused by the lead error, because it is a regular deviation along the helicoidal surface. By investigating the errors in the x, y, z direction for the theoretical and actual contact points, this mistake can be overcome.

The previously corrected measurement points (p_{zv} and the x', y', z' according to Figure 6.13) can be used to evaluate a helicoidal surface by mathematical calculation.

It would be useful to work out standards for three-dimensional surfaces. As the standards in use have prescribed characteristics, they

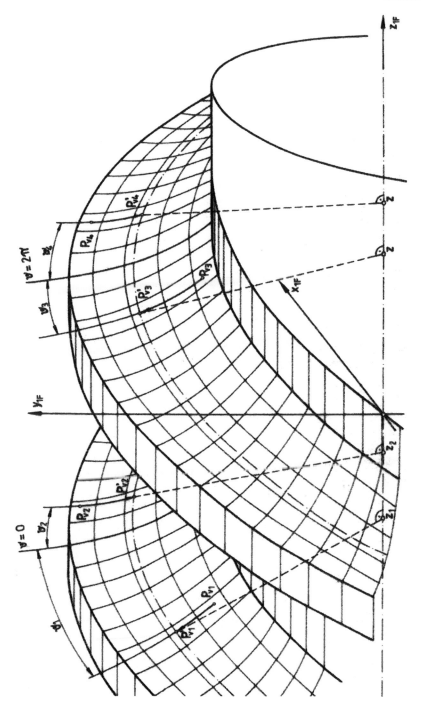

Figure 6.11 *Transformation of points measured on reference cylinder*

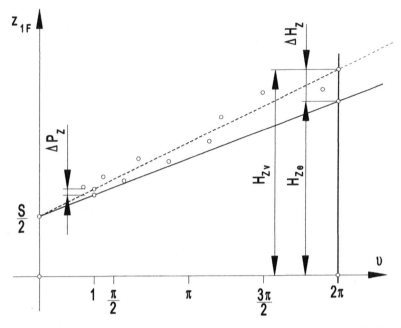

Figure 6.12 *Determination of the leads, the theoretical, p_{ze}, and calculated, p_{vz}, based on measurement*

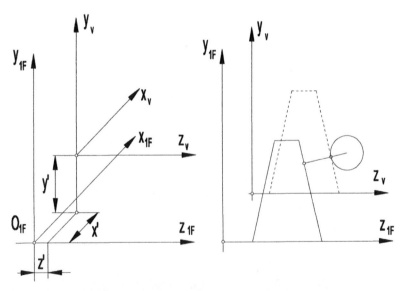

Figure 6.13 *Correlation between coordinate systems; theoretical helicoidal surface and one chosen for measurement*

have to be determined in different ways and so, to evaluate them, the points measured have to be transferred in different ways.

To evaluate the axial pitch and axial section, the profile contact points should be transferred into the y_{1F}-z_{1F} plane along the helicoidal surface.

The value of radial runout should be determined in plane x_{1F}, y_{1F}. The deviation of the thread curve on the helicoid surface is defined in the reference cylinder so it should be transformed there.

The theoretical block scheme of the 3D measurement and calculation is shown in Figure 6.14. The outputted program can present all the characteristics of the measurement program that traditional measurements provided before. An example is given in Figure 6.18.

The use of 3D coordinate measuring devices make it possible for the surfaces ground by the wheels profiled from calculated values to be assessed from a geometric point of view. During machining, the deviations caused by the deformations of the workpiece, of the device, of the machine, and of the tool (WFMT system) (Magyar, 1960) as well as the changes thus caused, can be numerically characterized so that the wheel-profiling CNC program can be corrected to obtain a wheel profile suitable for machine finishing of minimum error.

6.4 RESULTS OF MEASUREMENT OF HELICOIDAL SURFACES

The most frequently used and traditional method of measurement and of assessment of helicoidal surfaces in workshop practice employs the Klingelnberg PSR type measuring equipment, as shown for a cylindrical arched profile worm in Figure 6.5 and for a spiroid worm in Figure 6.15.

The measurement using an up-to-date 3D measuring device (Mitutoyo B320 type) is illustrated in Figure 6.16. It can be seen that the worm without any positioning is situated on the measuring table. The two V-blocks keep the worm safe from displacement.

The data measured and calculated for the worm in Figure 6.16, together with their evaluation, are tabulated in Figure 6.18.

From the data the effects of the applied corrections can be deduced. The seven qualifying characteristics (using several points selected) are shown opposite the traditional measuring method where each characteristic is determined by separate measurement.

The example presented is an Archimedian worm measurement

but the method worked out can be used for assessing other types of helicoidal surfaces as well.

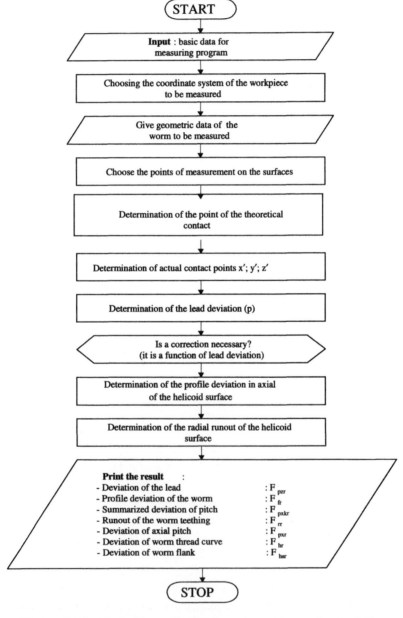

Figure 6.14 *Block schematic of 3D measurement and calculation*

Figure 6.15 _Checking of spiroid worm using Klingelnberg PS 12 type measuring device_

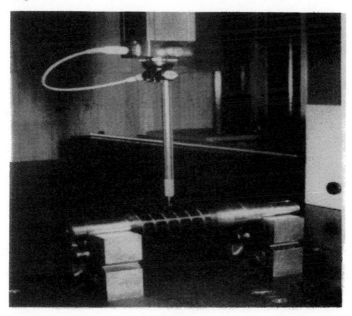

Figure 6.16 _Checking of arched profile worm using Mitutoyo measuring device (without circular table)_

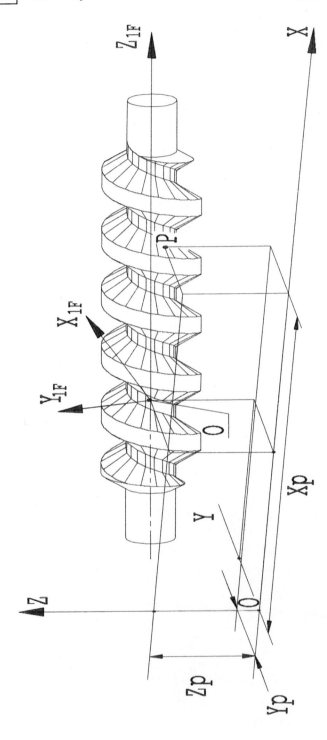

Figure 6.17 *Choosing the worm coordinate system on a 3D measuring device*

1.Thread Results Of The First Measurement (No Correction Applied)				1.Thread Results Of The Final Measurement (Corrected)			
FEELER	ACTUAL	THEORETICAL	DIFFERENCE	FEELER	ACTUAL	THEORETICAL	DIFFERENCE
Y -27.046	-27.06	-27.046	-.015	Y 20.40	20.42	20.412	8E-03
Z 26.969	26.978	26.968	.009	Z 28.697	28.713	28.7	.012
X 41.544	41.498	39.958	1.539	X -3.431	-3.377	-3.373	-5E-03
Y -31.922	-31.938	-31.922	-.017	Y -26.694	-26.71	-26.694	-.017
Z 31.142	21.148	21.141	.006	Z 19.227	19.238	19.229	8E-03
X -.442	-.488	-2.04	1.552	X .979	1.03	.992	.037
Y -26.691	-26.709	-26.691	-.018	Y -31.925	-31.939	-31.925	-.015
Z 19.233	19.241	19.232	.008	Z 21.136	21.145	21.14	SE-03
X 1.43	1.376	-.167	1.543	X -.893	-.835	-.882	.047
Y -1.17	-1.174	-1.17	-.005	Y 12.738	12.744	12.739	4E-03
Z 39.354	39.37	39.353	.017	Z 33.484	33.502	33.488	.013
X 11.115	11.07	9.551	1.519	X 11.049	11.104	11.114	-.011
Y 11.379	11.381	11.379	.002	Y 11.376	11.381	11.377	4E-03
Z -34.373	-34.388	-34.374	-.014	Z 34.018	34.036	34.02	.016
X -13.945	-13.955	-13.946	-.009	X 11.124	11.179	11.19	-.011
Y -34.373	-34.388	-34.374	-.014	Y -1.173	1.174	1.172	-3E-03
Z -13.945	-13.975	-13.946	-9E-03	Z 39.348	39.365	39.35	.015
X 30.436	30.388	28.852	1.535	X 10.664	10.724	10.712	.011
Y -25.697	-25.715	-25.698	-.018	Y -23.025	-23.036	-23.024	-.0013
Z 21.024	20.033	21.024	.009	Z 28.755	28.769	28.759	9E-13
X 43.579	43.526	41.996	1.529	X 27.189	27.246	27.221	.024
Y 23.847	23.857	23.847	.009	Y -34.376	-34.392	-34.374	-.019
Z 24.193	24.208	24.193	.014	Z -13.951	-13.957	-13.948	-9E-03
X 67.76	67.708	66.248	1.46	X 29.985	30.042	30.011	031
Y 15.628	15.633	15.628	.005	Y 15.625	15.633	15.627	6E-03
Z 31.141	31.158	31.141	.016	Z 30.135	31.152	31.137	.015
X .39.903	39.852	38.361	1.49	X 39.452	39.506	39.52	-.014
Y -23.022	-23.036	-23.022	-.014	Y -27.049	-27.061	-27.047	-.014
Z 20.761	28.772	28.76	.012	Z 26.963	26.975	26.967	8E-03
X 27.64	27.592	26.062	1.53	X 41.093	41.151	41.118	.032
Y 12.741	12.744	12.74	.004	Y -25.7	-25.715	-25.698	-.018
Z 33.49	33.507	33.489	.018	Z 21.018	21.03	21.02	9E-03
X 11.5	11.451	9.955	1.496	X 43.128	43.179	43.156	.023
Y 20.413	20.42	20.412	.008	Y 23.844	23.857	23.846	.01
Z 28.703	28.719	28.702	.016	Z 24.187	24.202	24.189	.012
X -2.98	-3.03	-4.533	1.502	X 67.309	67.362	67.409	-.048

Coordinates of the axial section

Radius (Y, 1F)	Z, 2F
35.234	-1.986
32.917	-1.099
38.304	-3.053
35.844	-2.215
35.889	-2.232
39.383	-3.484
36.855	-2.547
37.115	-2.634
34.855	-1.859
33.219	-3.031
33.220	-1.224
33.984	-1.573

1. Deviation of the lead	F_1PZR	: 12.3713E-03
2. Run out of worm teething	F_1RR	: 1.402E-02
3. Deviation of axial pitch	F_1PXR	: 12.3703E-03
4. Profile deviation of the worm	F_1FR	: 2.315E-02
5. Devn of the worm tooth flank	F_1HSR	: .0499900176
6. Devn of the worm thread curve	F_1HR	: 2.024E-02
7.Devn of summmarised pitch on worm	F_1PXKR	: 4.1109E-02

Data of the measured worm:

Teeth N°	: 1
Module	: 4.5 (MM)
Tool pressure angle	: 20 (FOK)
Diameter of the reference circle	: 72.6 (MM)

Figure 6.18 *Result of checking Archimedian worm using Mitutoyo measuring device*

7

MANUFACTURE OF HELICOIDAL SURFACES IN MODERN INTELLIGENT INTEGRATED SYSTEMS

At each stage in the process of production of worm gear drives, during design, manufacture and assembly, faults can occur. Modern intelligent integrated systems (ISS) (Su and Jambunatthan, 1994) can handle manufacturing in a versatile and flexible way; they can be efficiently utilized both in design and at the different phases of manufacture to improve product quality. Artificial intelligence and expert systems can now be used in the production of worm gear drives.

7.1 APPLICATION OF EXPERT SYSTEMS TO THE MANUFACTURE OF HELICOIDAL SURFACES

In the early phases of industrial development (up to the 19th century) machines and their components were manufactured, assembled, checked and assessed to a standard determined by the state of technology at the time. Present-day technology can provide more precision in design and manufacturing methods and the necessary tools and measuring devices.

Traditional manufacture relied on the engineer or craftsman who knew both the construction and the way of manufacturing the product (the technology) as well as the requirements of the product itself. The checking of products was based on professional experience, by touch and by use of operational tests. So from engineering feedback from the workpiece, corrections were generated manually and used to improve the manufacturing process.

Nowadays computer-aided design and manufacture using CNC machines, and checking and assessment using CNC devices, provides the possibility of correction at an advanced level given proper networking.

Today, intelligent CIM systems (dynamical systems) and advanced level automation (the process), including the possible use of computers, allow the use of feedback modelling into what previously was the domain of human intelligence.

The advances in artificial intelligence make possible the application of expert systems to the design and manufacture of helicoidal drives.

Expert systems are one of the central elements in modern intelligent integrated systems making it possible to help the human operator direct the process, and in some cases to replace him.

In this chapter we review modern intelligent integrated systems and the possibilities for application of expert systems to the design and manufacture of helicoidal surfaces.

7.1.1 Problems of manufacturing worm gear drives

Grinding guarantees high gear efficiency, high load-carrying capacity and low noise of operation of worm gear drives. This is achieved by fine tolerance control leading to near perfect running conditions and good-quality gear tooth surface finish.

The basic problem is to define the identity of surfaces enveloped with cutting edges of generating milling cutters for worms and worm gears (Figure 3.41). The basic problem in the grinding of helicoidal surfaces is created by the change of diameter of the gear teeth which is directly related to wear of the grinding wheel, so to generate the same helicoidal surface another wheel profile is required. The other important source of error is the mis-adjustment of tools used both for worm and worm gear machining. The geometric checking of both elements using CNC measuring devices and the evaluation of the results of measurement make it possible to detect the origin of the mistakes or errors and to feed them back into the manufacturing process.

Expert systems evaluate these deviations in an intelligent integrated system (ISS) having its own appropriate subroutine database.

This is advantageous, especially in the case of conical worms (spiroids) because to produce exact finishing of conical helicoids in an axial direction point-by-point, different grinding wheel profiles would be necessary (Figure 3.33).

7.1.2 Structure of the system

In previous chapters we have reviewed a method suitable for the design, manufacture and assessment of helicoid drives from a geometric precision point of view, and how it can specify the basis to build up different types of software for an expert system. The essence of this can help to create an expert system.

The following activities should be integrated into the system:

- conceptual design: the design process itself, results of specifications, etc;
- detailed design: analysis, choice of engineering material, drawings, etc;
- manufacture: tools, CNC programming, etc;
- checking, measurements.

These elements can be classified into five grades (the basic functional units):

- knowledge base;
- numerical calculations;
- databases, data handling;
- graphical presentation, drawing;
- control of all devices used (machines driving units, etc).

The knowledge base is handled and used by the expert system and the associated neural network. Its structure is shown in Figure 7.1. The expert system coordinates the other elements.

7.1.3 The full process

The intelligent integrated system (for worm gear drives) provides the following:

- design specifications (conceptual design): the user has to give the specifications for input shaft speed velocity, kinematic

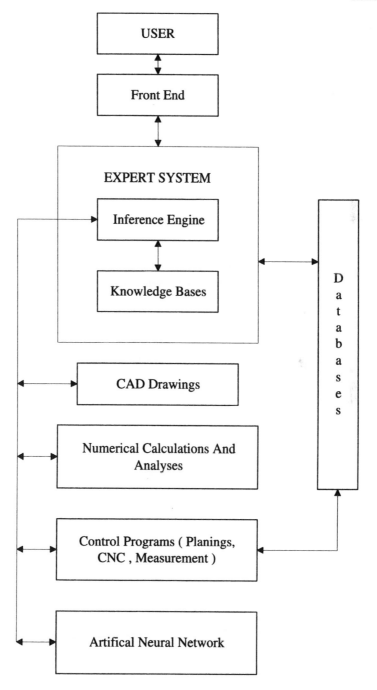

Figure 7.1 *Structure of expert system for design and manufacture of worm gear drives*

gearing ratio, power transmitted, relative position of shafts, centre distance etc. Using the above specification, the program offers different variations: module, teeth numbers etc;

■ detailed design: choice of engineering material (from database), determination of basic geometric data (Figure 7.2) and stress-state analysis (applying finite elements method) (Figure 7.12), workshop drawings of worm and of worm gear;

■ checking of documentation: when any result is not satisfactory it should be modified;

■ preparing CAD drawings: the CAD drawings of worm and of worm gear drawn from calculated values; the drawings should be saved in database;

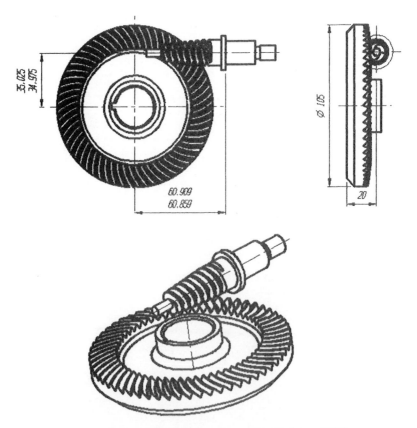

Figure 7.2 *CAD drawing of a spiroid drive (GM)*
I = 30, m = 1.33 mm, a = 35 mm, z_1 = 2, z_2 = 60

■ manufacture: selection of tools for worm and for worm gear machining, generation of CNC program, determination of grinding wheel profile (generation method of the CNC program), other data (eg data of the necessary adjustments for worm gear milling);

■ measurements: checking with CNC device the surface determined using basic data, analysis of the deviations and feeding them back into manufacturing process (eg wheel profile, machine and tool adjustments, etc).

7.2 INTELLIGENT AUTOMATION OF DESIGN AND MANUFACTURE OF WORM GEAR DRIVES

The design and manufacture of high-power transmitting worm gear drives have always raised problems. Subsequent to the development of spur cylindrical gear drives, there has been an increasing demand for different types of drive with:

■ kinematic drives (both for intersecting and for skew axes);
■ ever-increasing transmission power capability;
■ reduction of noise level between the interacting surfaces.

Some of these requirements demand changes in construction and others, in manufacturing technology used to lubricate gears.

The above-mentioned operational characteristics (geometry, precise manufacture and assembly) are closely related to each other and engineers, both designers and production planning specialists, require a high level of expertise. Much of this expertise is realized in the actual manufacturing process (eg worm grinding) and assembly (eg adjusting the proximity of the active contact surface) because most of the required operational characteristics can be attained during manufacture and assembly (see Chapter 8).

Because of the special character of the manufacturing of pared driving elements, a special, individual machine tool has been developed by machine tool manufacturers for each type of gear form to be generated. This is because universally applicable machines in this field cannot be used (eg Oerlikon, Gleason bevel gear generators, the Pfauter worm gear cutter, the Klingelnberg grinder for helicoidal surfaces, etc).

Complicated design and manufacturing are the reasons that this field (manufacturing elements of worm gear drives in intelligent systems) was retarded in spite of the fact that all machine tools generating different kinds of teeth are automated.

For these reasons, industrial and production design and, all the other functions, are best carried out in the framework of intelligent automation systems (Figure 7.3) (Dudás and Bányai, 1994; Su and Dudás, 1996) .

The field of application of intelligent automation is based on the control of automated processes (visual identification, measurement, analysis, inching, etc). Therefore, the use in manufacturing of intelligent devices is necessary to effect communication between the various process elements.

Although it is impossible to separate off parts of the complete design process, a few elements of it should be discussed from the point of view of creating intelligent automation.

Within the complete design process we can include:

- marketing;
- specification of industrial design;
- conceptual design;
- detailed design;
- manufacturing;
- sales.

In the following section, certain elements of conceptual design and manufacturing will be discussed. These elements will be discussed only with regard to cylindrical and conical worm drives. More detailed information on expert systems can be found in the literature (Breitung, 1993).

In conceptual design, the geometric characteristics first investigated are those that basically determine the actual contact area that primarily influences operational characteristics as well as the possibility of simulation of the operation. In manufacture, the finishing operation in machining which guarantees the quality of the product, ie the grinding, will be investigated as well as the quality control during manufacture of the product; all of this from the point of view of intelligent automation.

7.2.1 Conceptual design of helicoidal driving mates

7.2.1.1 Determination of the geometric characteristics influencing the actual contact area

The operational characteristics of worm gear drives are influenced by, among many other factors, the oil film wedge developed between contact surfaces. The manufacturing process should ensure

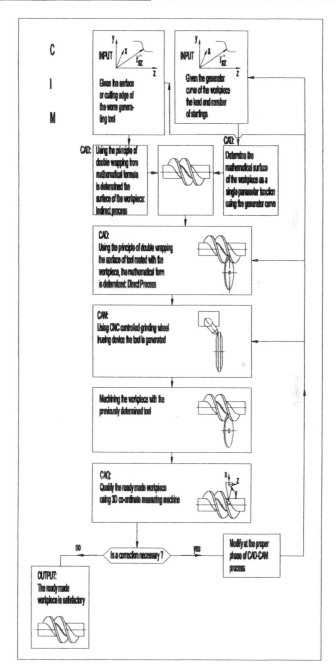

Figure 7.3 _Production of helicoidal surfaces in CIM system (Dudás, 1988a, 1988b)_
(University of Miskolc, Department of Production Engineering)

that such contact surfaces conform to being within the zone of mesh; characterized as when 40 per cent of the contact curve inclines 70°–90° to the direction of the relative velocities (Figure 7.5). These surfaces have a high load-carrying capacity and are able to provide suitable hydrodynamic conditions and sometimes hydroelastic conditions for oil lubrication. To improve contact conditions and to increase load-carrying capacity, one possibility is to keep the actual contact area within the appropriate range (Dudás *et al,* 1983). One way of reaching this condition is to choose geometric data so that they ensure existence of a localized 'actual contact area'. Coordinated geometric sizes are closely connected to the operational characteristics of the worm gear drive.

The different operational characteristics (efficiency, noise level, operational temperature, power transferred, service life, etc) of all drives are basically determined by local gear tooth contact conditions. Of the different meshing conditions the following should be mentioned (they are characteristic of mating elements):

- total length of contact curves;
- position and geometric shape of the contact curves;
- the relative angular difference between contact curves and relative velocity;
- the position of contact area.

Though the type of mating elements (the geometry) determines the limits of these characteristics, by optimizing the parameters, the drive can be maximized from several points of view. It should be noted that determination of the above-mentioned characteristics by computer calculation can only be reached by keeping the parameters constant (within the limits of the investigated range, the chosen steps and other uninvestigated parameters) to obtain results useful for the assessment of different solutions.

The equation of the contact curve according to the law of meshing is:

$$f(u, \varphi_1) = \bar{n}_{1F} \cdot \bar{v}_{1F} = 0 \tag{7.1}$$

For given values of the movement parameter φ_1, the parameter couples $\eta(u)$ and $\vartheta(u)$ conform to each other and fulfilling equation (7.1) can be determined.

So the contact curves characterizing the mesh are available and represent different characters using different parameters.

The drive is assessed in the light of the following three characteristics:

1. Total length of co-existing contact curves

The total length of co-existing contact curves, from equation (7.1), is:

$$L = \sum_{k=1}^{m} \int_{u_{jr}}^{u_{jt}} ds \tag{7.2}$$

where:

$ds = \sqrt{(\dot{x}(u))^2 + (\dot{y}(u))^2 + (\dot{z}(u))^2}$,
k is the running suffix of the contact curves,
m is the number of the investigated contact curves,
u_{jr} is the parameter u belonging to dedendum cylinder,
u_{jt} is the parameter u belonging to addendum cylinder.

The chosen criterion, eg load-carrying capacity, fixes the maximum of the equation (7.1):

$$\left. L_0 = \max_{j} \left\{ L_j(\eta, \vartheta) \right\} \right\} \tag{7.3}$$

where:

L_j is the values of L at different investigated versions,
L_0 is the optimum of L,
j is the running suffix of the investigated version (j = 1,2,...z),
z is the number of the investigated versions.

It should be noted that this value provides the point for evaluation at the optimal solution according to point b.

2. Relative positions of contact curves and relative velocities

The most advantageous situation from this point of view is when the angle of inclination between the relative velocity vector ($\vec{v}_{1F}$) and the tangent of the contact curve (E) is the closest possible to 90°:

$$H_z = \frac{1}{L} \sum_{k} \sum_{i} v_{1Fj}^{(12)} \cdot \vec{t}_i \tag{7.4}$$

where:

v_{1F} is the relative velocity vector,

i is the suffix of the investigated points along one contact curve,

k is the suffix of the investigated contact curve;

H_z is the values of H at different investigated versions.

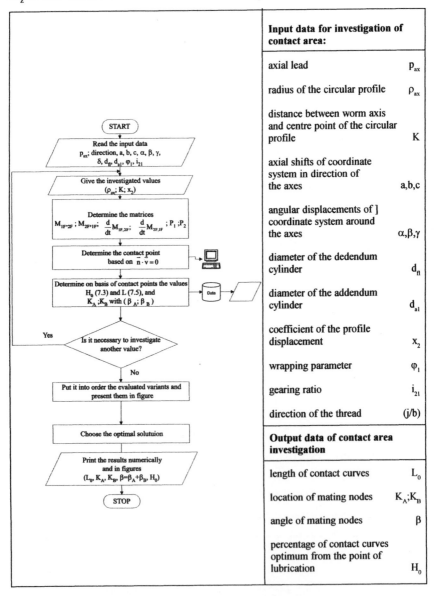

	Input data for investigation of contact area:	
	axial lead	p_{ax}
	radius of the circular profile	ρ_{ax}
	distance between worm axis and centre point of the circular profile	K
	axial shifts of coordinate system in direction of the axes	a,b,c
	angular displacements of] coordinate system around the axes	α, β, γ
	diameter of the dedendum cylinder	d_{f1}
	diameter of the addendum cylinder	d_{a1}
	coefficient of the profile displacement	x_2
	wrapping parameter	φ_1
	gearing ratio	i_{21}
	direction of the thread	(j/b)
	Output data of contact area investigation	
	length of contact curves	L_0
	location of mating nodes	$K_A ; K_B$
	angle of mating nodes	β
	percentage of contact curves optimum from the point of lubrication	H_0

Figure 7.4 *Determination of contact curve*

To evaluate, search for the most advantageous value:

$$H_0 = \min_{j}\left\{H_j\left(\eta,\vartheta\right)\right\} \tag{7.5}$$

where:

z is the number of the investigated versions,
H_j is the values of H at different investigated versions,
H_o is the optimum for H,
k is the suffix of the investigated contact curves,
j is the running suffix of the investigated contact curves (j = 1,2,...z).

This in an important characteristic expressing the formation and load-carrying capacity of the oil film.

3. Arrangement of the contact curves

The third evaluation serves to modify the maximum and minimum values obtained from consideration of the previous two points. In this evaluation we assess the arrangement of the contact curves and the position of the meshing modes from the point of view of the efficiency of the drive. This method, according to the block schematic shown in Figure 7.4, can be used to determine, for a given type of drive, those optimum geometric parameters that guarantee the ideal position of the contact area (Figure 7.5).

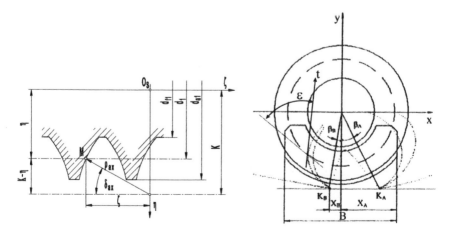

Figure 7.5a *Interpretation of the effect of geometric parameters on contact curves and their positions (for ZTA worm gear drive)*

The effect of geometric parameter changes on the position of contact curves, shown in Figure 7.5, proves the assumption in the case of the example shown.

For running in, the value of percentage $\alpha_{jó}$ represents the percentage of those values that fall in the range $\varepsilon = 70°-90°$ of the angle e belonging to all the points investigated.

The effect of the increase of the profile displacement coefficient (x_2) is to:

- significantly reduce the total length of the contact curves;
- reduce the distance x_B (the distance x_A too);
- reduce the angle ß;
- increase the angle ε.

The effect of the increase of ρ_{ax} is to:

- have no significant effect on the total length of contact curves;
- affect x_B as a function of the profile displacement coefficient (x_2).

The effect of the increase of the value K is to:

- reduce total length of the contact curves;
- generally reduce the length x_B;
- significantly reduce the angle ß;
- increase the angle ε as consequently lubrication is worse?

In a worm designed and ground in the way described above, such a helicoidal surface will be generated that produces a common point (K_B) of the contact network which is positioned along the line connected, at about 1/6 B from the main contact point C (B is the face width of the worm gear). In this case, the so-called lubricating wedge between the mating surfaces is formed as well as the limited field for teeth contact. The range for profile displacement guaranteed by the above condition is $x_z = 0.8 \div 1.5$.

Using this method, worm gear drives of higher efficiency than similar cylindrical drives manufactured by the traditional method can be obtained.

Using this program, worm gear drives having any arbitrary parameter, given the same conditions, could be evaluated, and the parameters, necessary to optimal meshing, can be chosen.

By harmonization and optimization of the radius of the surface curve (ρ_{ax}), the profile displacement coefficient (x_2) and the

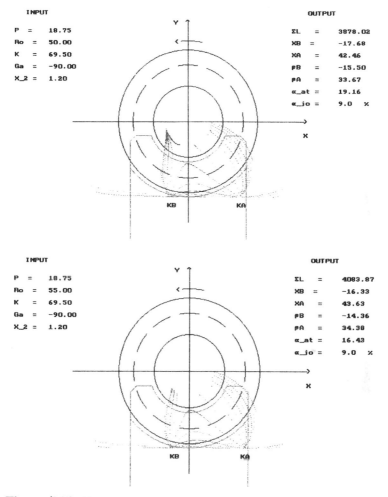

Figure 7.5b *The effect of changes in* ρ_{ax} *(ZTA worm gear drive)*

position of the centre of surface curve (K), a better and more limited contact field can be obtained.

The literature (Krivenko, 1967) supports well the result obtained, namely that the distance between K_A and K_B and their positions can be guaranteed by suitable selection of geometric parameters, that is:

$$\beta = \beta_A + \beta_B$$
$$\beta = 45 \div 50°$$
$$\alpha_{ax} = 20 \div 22°$$
$$\text{and } 1 \leq x_2 \leq 1.5$$

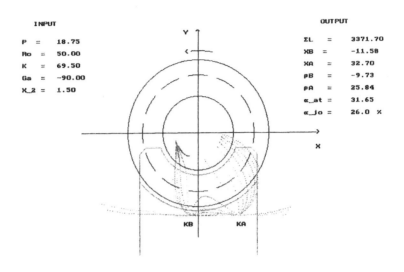

INPUT

P	=	18.75
Ro	=	50.00
K	=	69.50
Ga	=	-90.00
X_2	=	1.50

OUTPUT

ΣL	=	3371.70
XB	=	-11.58
XA	=	32.70
ρB	=	-9.73
ρA	=	25.84
α_at	=	31.65
α_jo	=	26.0 %

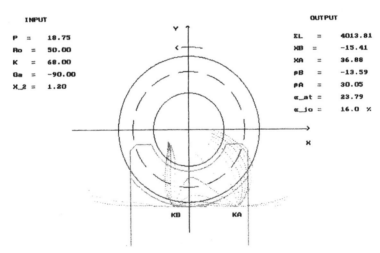

INPUT

P	=	18.75
Ro	=	50.00
K	=	68.00
Ga	=	-90.00
X_2	=	1.20

OUTPUT

ΣL	=	4013.81
XB	=	-15.41
XA	=	36.88
ρB	=	-13.59
ρA	=	30.05
α_at	=	23.79
α_jo	=	16.0 %

Figure 7.5c *The effect of changes in K (ZTA worm gear drive)*

In this section, the positions of contact curves of spiroid drives were examined. From Figure 7.5e it can be seen that in the case of definite mating of teeth, the contact curve is perpendicular to the relative velocity vector, so the spiroid drives are more efficient from the point of view of lubricating conditions as well.

7.2.1.2 Investigation by simulation of worm gear drive of given geometric parameters

During manufacture, the aim is to obtain results as near to the approximate theoretical values as possible. But practical conditions always differ from the theoretical ones and the deviations during operation change, the designed character and characteristics of meshing.

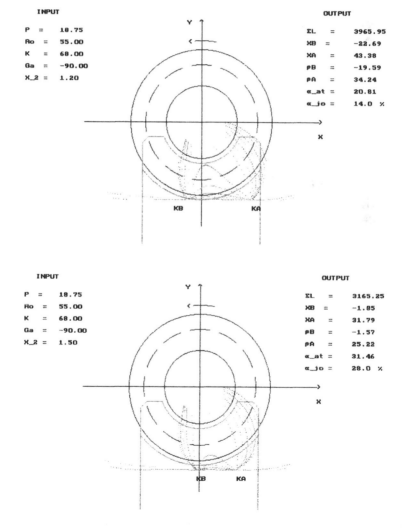

Figure 7.5d *The effect of changes in x_2 (case of ZTA worm gear drive)*

To make possible investigation and design of the actual meshing (without actually manufacturing and investigating a number of actually manufactured elements), the optimal values of the geometrical parameters and the actual mating conditions are determined in a theoretical way, taking into account the stochastic characteristics of the individual parameters (ie the surface conditions). These effects can be investigated by modelling or by simulation.

Geometric deviations can cause errors in the characteristics of the drive (position of the contact curves, shape, the position of the common points of the contact network at meshing, the extent and the position of the contact area, etc). The origin of geometric deviations can be classified as:

- manufacturing errors: all the errors occurring during manufacture, both with worm and worm gears, such as errors in tool size, in tool adjustment, errors due to kinematic deviations in the machine tool, the important parameters (eg lead of thread, the profile, the pitch, etc) show scattering round a nominal value within the stated tolerance;
- setting errors: manufacturing errors in the drive housing influence the meshing of the mated elements, because the relative position of worm and worm gear (centre distance, angle of inclination between axes, position of axial section, etc) can only be adjusted allowing limited error and the cylindrical worm gear drive is extremely sensitive to these mistakes (Figure 4.6);

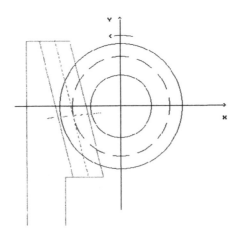

Figure 7.5e *Spiroid worm: position of contact curves*
(m = 5 mm, d_j = 64 mm, β_j = 10°, β_b = 30°)

- elastic deformations: the elastic deformation of the elements of the drive during power transmission compared with that in theoretical meshing causes geometric deviations in the vicinity of the meshing as neither the worm nor the worm gear are ideal rigid bodies (Figure 7.9);
- thermal load: though its extent is not significant, heat expansion causes deviations compared with the theoretical values due to differences in temperature between those during manufacturing and checking and those during operation; these can be significant when operating at high speed;
- surface roughness: can be regarded as a collection of micro-geometric deviations because surfaces are not ideally smooth and cause asperity contact instead of line contact.

It is impossible to appreciate the full actualities of the practical situation when operating an economic and easy-to-handle simulation program, so some flexibility is needed in the theoretical correlation to make the simulation suitable for handling.

The following have to be taken into account:

- the two surfaces envelop with each other (conjugated surfaces) and the worm and the hob tool for the worm gear are different because the generated surface differs, owing to manufacturing errors, from the surface that meshes with the worm; this fact must be taken into account when considering the laws of theoretical meshing;
- the surfaces are not ideal ones; the actual points of the surfaces show scatter;
- the contact curve does not follow a line contour but, in the vicinity of the loading is a more point-like contact and this point wanders continuously;
- the effect of the oil film during the investigation of the meshing is ignored as its thickness is only a few μm, equal in scale to the actual surface roughness; it is only of significance because the plastic deformations are also of this scale of dimension.

These conditions have to be applied for contact points in the vicinity of the contact curve, as was mentioned above. First of all, the contact curves should be determined for the circular profile in axial section (ZTA) of the worm gear drive, taking into consideration the variability of different parameters (Figure 7.8).

The mating worm and worm gear carry significant load. This load will spread the area of contact in an ellipse-like instead of a point-like area of contact. The measure of the actual contact ellipse depends on the principal curves of the mated surfaces and their orientation (Litvin, 1972).

The shape of the contact area depends in particular on the contact curves over the tooth flank, their position and dimensional area. The above characteristics are basically influenced by the design of the constructional and the machine-adjustment parameters, together with assembly error (Figure 7.8).

With regard to contact problems, it is necessary to generally determine the following:

- position of contact domain;
- distribution of contact pressure;
- the ideal 'rigid like' approach of teeth contact.

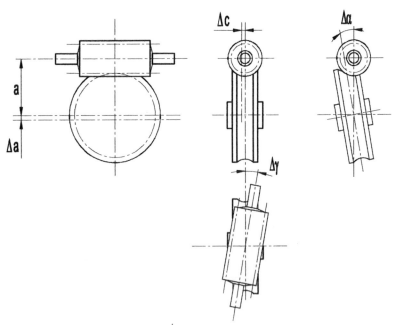

Figure 7.6 *Assembly of worm gear drives; typical mistakes (a – centre distance, Δa – deviation of centre distance, Δc – error of position of the worm axis, $\Delta \alpha$ – error in angle of rotation of the worm gear, $\Delta \gamma$ – angle error of adjustment of the worm)*

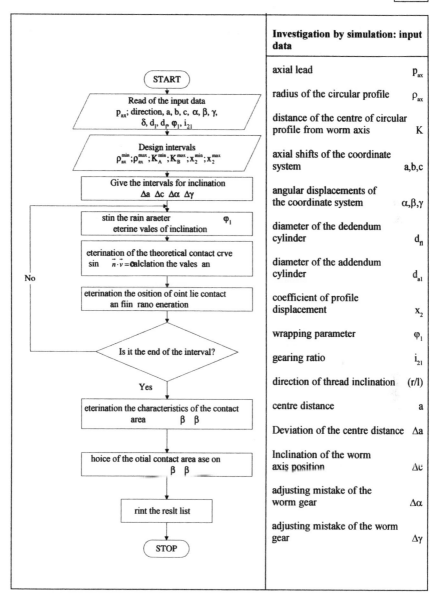

Figure 7.7 *Investigation by simulation*

There are several theories relating to contact conditions, and of them, the Hertz theory contains too many simplifying approximations so is only suitable for analysis of simpler problems. These days, with the development of computer techniques, numerical methods which employ significantly fewer simplifying assumptions are available to solve more complicated problems. The appropriate method is solution by finite elements, whose application to solve contact problems will now be considered.

The Hertz theory is the basis for several complicated contact theories, some interactive, others using the theory of variables. Another applies an effect-matrix to determine the actual contact conditions.

Returning to the actual contact area in worm gear drives, it should be noted that it plays a key role in correct operation and in load-carrying capacity. The principle is summarized and results shown for the new procedure for the evaluation of the stress analysis in Figures 7.11 and 7.12.

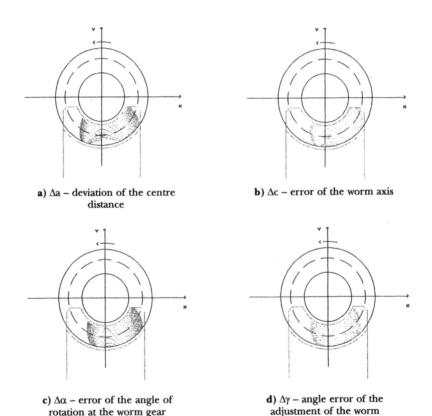

a) Δa – deviation of the centre distance

b) Δc – error of the worm axis

c) Δα – error of the angle of rotation at the worm gear

d) Δγ – angle error of the adjustment of the worm

Figure 7.8 *Effect of adjustment of mistakes on bearing pattern (ZTA worm gear drive)*

For the determination of the contact curves belonging to the flanks of worm and worm gear for a given angular displacement φ_1, a computer program based on the kinematic method has been developed. The program for a given worm geometry calculates the theoretical contact curves and fits on them a envelope surface. This enveloping surface, after it is transferred into a system joined to the worm gear, makes it possible to generate the geometric model of the mated elements suitable for a finite elements program. Further, it determines the direction of the contact pressure at the points of the contact curve that are necessary to calculate the load perpendicular to the tooth flank.

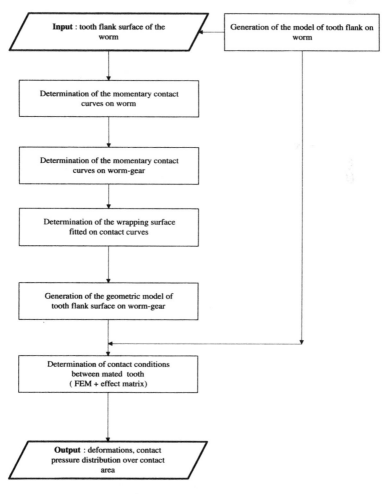

Figure 7.9 *Block schematic of FEM*

Figure 7.10 *3D image of GM spiroid drive*

The aim was to simulate the loads during meshing to evaluate the damage, the site and value of the deformation as well as to study the stress state.

Using the I-DEAS computer program, a corresponding module was used to carry out the finite element investigations.

The meshing elements (worm and worm gear) were evaluated, following Boolean algebraic rules in the 3D system.

The figures reproduced represent the simple case when the worm is clamped in an ideal rigid way and a single concentrated force is the load. It is rare in practice for a generally distributed load to be taken into account (Figure 7.11) (Simon, 1996).

Setting the clamping and the load, the system generates automatically the finite element mesh by building in the chosen spatial element on the basis of the adjusted values. Arched spatial elements were used in the example.

Figure 7.12, which emerged after post-processing, provides information on site and value of the deformation; the stress state can be judged by the scale on the right.

The numerical results of I-DEAS system calculations are displayed on the upper left part of Figure 7.12.

7.2.2 Manufacture of worms and worm gears

7.2.2.1 Structure of the manufacturing system

The function and correlation of the elements that take part in the manufacturing process are presented here as if they would constitute an elastic manufacturing cell and, regardless of their connection and function with the other elements of the system, manufacture and evaluation through measurements of helicoid surfaces only will be discussed. The structure of the system can be seen in Figure 7.13.

The functions of the elements with regard to their informational relations are as follows:

1. Central computer:

■ to generate the equation of the theoretical surface using the actual parameters of the mathematical model;
■ to carry out preliminary data processing for control of the thread grinding machine tool, control of the wheel-truing device, control of 3D measurement unit, to provide data for CCD camera as another type of measuring unit;
■ to optimize movement of the thread grinding machine tool and the surface generator milling machine tool by minimizing the truing of grinding wheel;
■ to process data obtained from the 3D measuring machine and measuring unit which, in case of need, carries out corrections to control machine tools for grinding and for generating;
■ to carry out necessary documentation.

2. Control unit of the thread grinding:

■ using data obtained from central computer, it controls movements of the grinding machine and controlling device;
■ to compare information obtained from wear checking on the grinding wheel and the data served by central computer and, in case of errors, achieves correction.

3. Control unit for measuring machine:

■ control the stylus at necessary points using data obtained from the central computer (the theoretical surface);
■ after preliminary data processing, results of measurements will be provided for the central computer.

These main functions in important cases require application of intelligent devices. Naturally, in the preliminary data processing carried out by the central computer, the processing of results can in part be transferred towards the other control units provided their intelligence is adequate.

7.2.2.2 Grinding of helicoid surfaces

Basically, two problems arise with the finishing by grinding wheel of a conical helicoid surface, which make the formation of the geometrically correct profile practically impossible. These are:

- the wear of the wheel during grinding causes change in both the profile and the diameter of the wheel; this is the reason why the worm profile becomes distorted compared to the theoretical profile and even to the original profile at the start of finishing (see Chapter 4);
- the change of the diameter along the axial section of the conical helicoid surface itself causes the continuous change in the worm profile (in case of constant wheel profile) (see Chapter 3).

The first problem occurs in practice during grinding of cylindrical helicoid surfaces as the worm profile depends on the centre distance of the worm and the wheel in the case of constant wheel profile (Figure 4.8), and precise grinding can only be obtained by using a superhard grinding wheel or by frequent wheel dressing and the result of its being fed back from the worm.

The second problem is generated by the geometry of the conical helicoidal surface itself and depends on both the centre distance and the changes of diameters of the worm and the grinding wheel. This causes a change in ratio of the diameters, which is independent of wear of the wheel, along the centre line of the worm, resulting in step-by-step profile distortion (Figure 7.14).

The possible benefits of conical worm grinding vary from exact machining geometry to economic working, appropriate to practical requirements. In practice, compromises are made to find appropriate solutions for a given task (see Chapter 5).

The following three possibilities present themselves. They have already been discussed in previous chapters, but it is now necessary to reintroduce them, from the point of view of operation of the system as a whole.

Figure 7.11 *Finite element mesh: clamping*

1. Continuous truing of the wheel

The worm and the grinding wheel, when in relative motion, envelop
each other. The conical worm surface needs to be guaranteed when
there is a continuous change of the wheel profile. From Figure 7.14
it can be seen that, during the wheel movement from position 1 to
position 2, continuous correction is necessary in order to achieve
a 'distortion free' wheel profile, owing to continuous changes of the

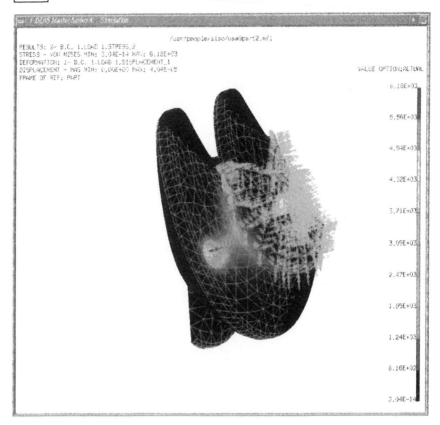

Figure 7.12 *Final result of the finite element analysis*

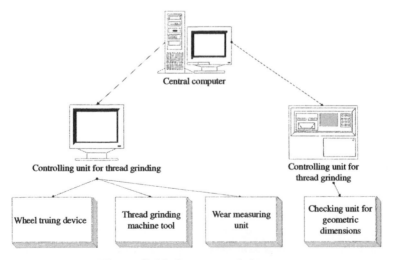

Figure 7.13 *Structure of the system*

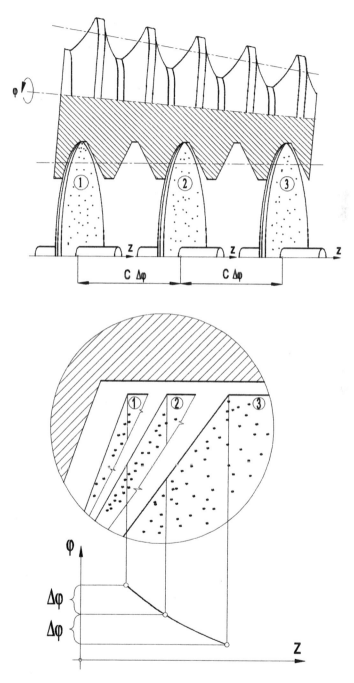

Figure 7.14 *The changes necessary of wheel profile as a function of conical worm diameter for a given conical worm profile*

centre distance and ratio of diameters a. To achieve this, a program package for a CNC-controlled wheel-regulating device is required, which is able to bring about truing of the wheel on the actual profile necessary for the axial position of the wheel. This program package makes available the necessary mathematics for meshing and, therefore, a controlled wheel truing device (Dudás, 1988a).

The advantage of this method is that it provides theoretically precise machining; its disadvantage is that continuous truing causes substantial material wastage of the wheel material, enough to render the machining uneconomical. Another disadvantage of this method is that regulating the pencil-shaped tool generates a spiral surface on the grinding wheel, resulting in inaccuracy which varies with the shaft speed.

The sketch representing the operation of the CNC wheel-truing device can be seen in Figure 7.14.

2. Optimizing the wheel profile

This method provides the precision needed to meet practical requirements to keep the wear of the wheel to a minimum. This is what controls the wheel profile, so a precise worm profile at one place somewhere between the maximum and minimum diameters of the worm is generated, but the profile distortion at both ends remains within the prescribed tolerances. That is, the optimal value of the movement parameter (φ) is determined, which assigns to the tool profile used during machining the absolute value of the minimum deviation from the theoretical value for the whole worm surface. In practice, it means that the parameter value determines a medium wheel profile, between the two profiles assigned by the two extreme diameters of the conical worm. As change in wheel profile as a function of the movement parameter is not linear, the optimal value is not the mean diameter along the axis of the thread surface but is a value of φ_{1opt} fulfilling the requirement:

$$\frac{Z_{2F}(\phi_{1min}) + Z_{2F}(\phi_{1max})}{2} = Z_{2Fopt.} \tag{7.6}$$

This is presented in Figure 7.15. The advantage of the method is that it is easy to carry out and the wheel consumption is small.

It is a disadvantage that, depending on sizes of the worm, the machining does not always guarantee precision within the tolerance prescribed. When dividing the surface into several parts, a discontinuity can be experienced at the extremities.

3. Kinematical truing of the grinding wheel

This method brings together the advantages of the previously discussed two methods (precise machining, minimum wheel truing) but the truing process is very difficult and can only be carried out by application of an intelligent automated system.

The essence of this is that precise profile shaping is achieved not only by truing of the wheel profile but by kinematic adjustment of the wheel so that by moving the grinding wheel it may be positioned precisely enough from the point of view of enveloping the workpiece.

To realize this it is necessary to bring together the total capacity of the manufacturing system, ie:

- the facilities of the measuring unit regarding wheel profile;
- the momentary positions and the adjusted position of the CNC thread grinding machining;
- applying to the theory of meshing, the values that are monitored by a program segment, the points of the worm profile calculated using the central computer;
- the continuous readiness of the control unit of the wheel truing CNC facility.

7.3 MEASUREMENT AND CHECKING OF HELICOIDAL SURFACES IN AN INTELLIGENT SYSTEM

Recently, technological process design procedure is governed by use of variable manufacturing systems and within this CAD/CAM field, CAQ systems have gained ground as the need increases for quick, precise and automated manufacturing and checking of differently shaped components.

The manufacture of helicoidal surfaces within a variable manufacturing system makes it possible, since both the geometry and the applied technology depend on their dimensions, that they can be machined on the same machine tool, so representing a family of components.

But for manufacture of each helicoidal surface, even for the smallest dimensional error, separate grinding wheels are necessary, making every workpiece an individually manufactured one.

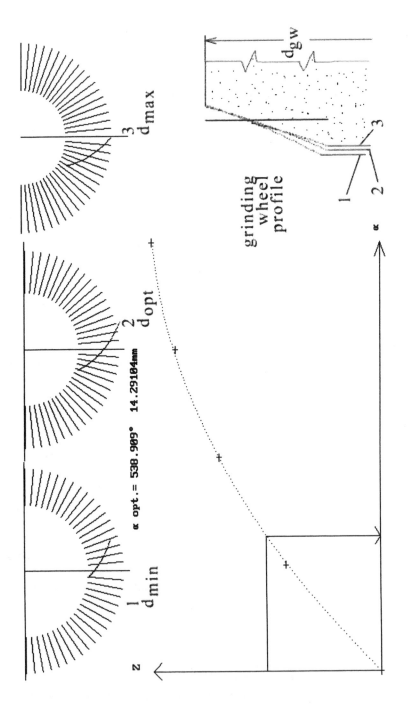

Figure 7.15 *Generation of the optimal wheel profile*

This problem can only be solved by detailed analysis of machining geometry.

The widespread adoption of the coordinate measurement method and its close connection to computer techniques requires that variable manufacturing systems, mainly used for machining helicoidal surfaces, should be developed further by carrying out geometric measurements and checking using CNC-controlled coordinate-measuring machines.

During the analysis of machining geometry, a mathematical model was built that is suitable for working out such a method of measurement (Chapter 6) which provides the possibility to check the geometry of arbitrary helicoidal surfaces in general using a three-coordinate measuring machine (Bányai, 1977; Lotze, Rauth and Ertl, 1996). The software system worked out on the basis of this measuring method provides the possibility to investigate the effect of the manufacturing parameters on shape deviation of the helicoidal surface as well as to analyse these effects and to facilitate the evaluation of the measurement results. It also makes possible the intervention into the manufacturing process of feedback. At the same time, the profile of machine tools for worm gears, using the profile points determined by the computer program, can easily be designed.

7.3.1 Checking of geometry using coordinate measuring machine

There are two methods available for the geometric checking of complicated spatial surfaces (either with or without direct contact):

1. The equation of the surface using a given coordinate system can be determined, and the coordinates of the measured point derived from the equation calculated, and then their deviations from the coordinates of the theoretical points have to be determined. Checking a few measured points can provide verification of the surface geometry using these methods.
2. When the surface is too complicated, ie it cannot be described by an equation in any coordinate system, this type of surface is represented theoretically by a set of points situated in a stochastic way in the majority of cases without a reference system. Then, the measurement involves surveying of the surface and, by fitting it on the theoretical surface, the deviations can

be determined. The surveying of the surface requires measurement of a large number of points and the treatment of the results is lengthy (eg surface regression).

The following basic requirements should be taken into consideration when preparing the measurement program:

- the time needed for the measurement, from beginning to the end of the evaluation of the results, should be less than the manufacturing timescale;
- precision of evaluation should be within the recommended manufacturing tolerance;
- the results of evaluation make possible the determination of the technological parameters missed.

Taking into consideration the above requirements for measurement of helicoidal surfaces, method 1. was chosen. The theoretical block schematic can be seen in Figure 6.14 (Bányai, 1977).

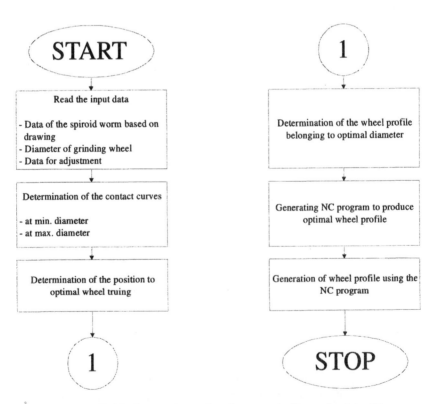

Figure 7.16 *Generation of optimum grinding wheel profile*

7.4 DEVELOPMENT OF THE UNIVERSAL THREAD-GRINDING MACHINE

There are many available methods for the grinding of thread surfaces. Depending on whether the threads are used as joining elements or whether they are kinematic ones, the dominant requirement being the dimension or precision, they differ from each other by a significant measure. The geometric requirements basically influence the manufacturing process.

This book primarily deals with the development of concepts of thread-grinding machines for the manufacture of kinematic thread surfaces of high precision.

This problem is a real one because, according to present knowledge, no thread-grinding machine tool commercially available is suitable for the universal finishing of the newly developed worm gear drives (eg spiroid, globoid).

Most up-to-date CNC-controlled thread-grinding machines are suitable for manufacturing a given type of thread surface within a well-defined range of dimensions. The target is to develop a thread-grinding machine suitable for universal adoption.

7.4.1 Review of thread surfaces from the point of view of thread-grinding machines

Previously mentioned thread surfaces with constant lead can be classified into three groups as:

- thread surfaces for joining elements;
- kinematic thread surfaces;
- tool surfaces.

Additional types of thread surfaces are not discussed in detail here (eg worms of extrusion). Here, the grinding of kinematic thread surfaces is discussed (cylindrical and conical worms, ball-thread spindles, feed spindles, etc). These require a high level of finishing precision. They are summed up in Figure 7.17.

7.4.2 Manufacturing problems of thread surfaces

The grinding of thread surfaces raises two basic problems: productivity (eg double conical and threaded grinding wheels etc) and precision (eg finger-like grinding tool etc).

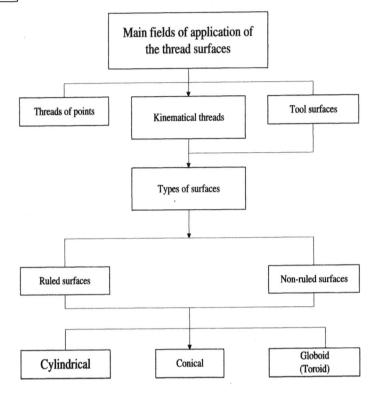

Figure 7.17 *Summary of thread surfaces*

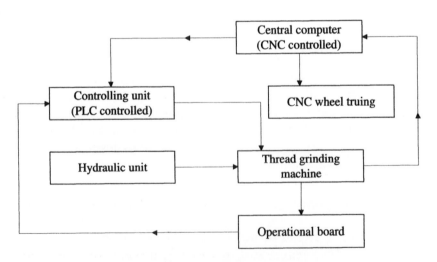

Figure 7.18 *The structure developed*

The two problems are not mutually exclusive, but they can be dealt with by the use of appropriate technology and machine tools.

The basic problem that arises when grinding thread surfaces is that the contact curve of the thread surface and a surface of rotation is a function of the respective diameters or a ratio of them. (Naturally the given profile should be taken into consideration.)

The problem is even more complicated in the case of a non-cylindrical surface, eg a conical (spiroid) worm that should be ground. In this case not only is the wear of grinding wheel manifested in a change in its diameter causing distortion of the thread surface (for constant profile), but the change in the conical worm diameter along its axis (change in lead angle) also requires changes of the grinding wheel profile from place to place (Figure 7.14). Here it should be noted that with the use of a conical grinding tool, the distortions are reduced but then the productivity is significantly less; therefore this solution is mainly applied in manufacture of the tool (eg the machining of the back surfaces of generating milling cutters).

At the same time, the tool necessary to produce the worm gear has to be manufactured. The grinding of the back and side surfaces of the hob cutter and the re-sharpening of its flank requires the solution of serious geometric problems too; otherwise, it has to be carried out on a machine tool. But traditional thread-grinding machine tools are not always suitable for exact manufacture of the tool. In the case of a conical worm hob, the grinding of the back of the side (or the back) surface with the necessary constant lead, the problem is more complicated because its teeth are positioned over a conical surface.

For the machining of a back surface generated by radial back-machining, it is most advantageous to use a CNC-equipped grinding machine connected to a wheel control appliance; when continuous wheel control is not used, the wheel can be optimized on the back surfaces of the teeth on the hob as part-surfaces of the domain of conical surfaces (Dudás, 1988c, 1991; Dudás, L., 1991).

7.4.3 Requirements of the thread-grinding machine

Regarding control requirements, there is expense involved both in the employment of the number of controlled shafts and in the expense of introducing control itself, which is increasingly progressive (eg the cost of controlling a five-shaft system can be much more than the cost of controlling 3+2 shafts).

But the cost of control can be reduced to a fraction of the original by application of correct function analysis.

Nevertheless, the machining, geometric and manufacturing requirements must be kept in mind.

The universal thread-grinding machine should be suitable for the machining of thread surface as shown in Figure 7.17. To achieve this, the following should be controlled:

- speed of the object spindle either as a function of coordinates or of time;
- the radial displacement of the grinding wheel housing in relation to the axis of the workpiece;
- horizontal swivelling of the grinding wheel head round a given axis;
- vertical displacement of the wheel truing device.

In the case of controlling five shafts, the types of thread surfaces shown in Figure 7.17 can be ground.

To grind a thread surface of varying lead, it is useful to control the movement of the object table lengthwise. For five-shaft control, the work can be divided into: control of three shafts for kinematic generation of the thread surface, and control of two shafts for truing of the grinding wheel profile.

All the other tasks that arise during generation of all the thread surfaces can be solved by use of simple control and operational commands.

7.4.4 Development of a possible version

To take into consideration the kinematic and control tasks enumerated above in the present state of development of the universal thread-grinding machine is the aim of our team. The structure of the project can be seen in Figure 7.18.

Control has been achieved with partial success by applying this structure, CNC control having provided fulfilment of basic functions or cycles.

The condition for exact, geometrically precise manufacture is a proper wheel profile, for which a licensed version is available, built in the form of a wheel truing CNC device technically equipped with proper software (Dudás, 1988b; Dudás and Csermely, 1993). This will not be discussed in detail here.

After realizing in practice the tasks involved in control problems, the universal thread-grinding machine will be suitable for machining to the exact geometry using the software system intended for control for all types of thread surfaces enumerated in Figure 7.17.

7.5 CONCLUSIONS

The intelligent automation of the manufacture of worm gear drives has come about because of the requirements of modern manufacturing. In the present state of manufacturing only some of these requirements have been fufilled because a comprehensive system has not been built at present.

In this chapter, the structure of an overall system was described as well as the considerations and methods relating to some elements of it. Some elements of the system (CNC wheel truing, measuring program and conceptual design programs) have already been realized. The working out of others (eg the thread-grinding machine) and incorporating them into the system is the most important theme for our team as a research activity. We estimate that both the theoretical bases and methods and the up-to-date available tools will improve both the productivity and the quality of products.

8

MAIN OPERATING CHARACTERISTICS AND QUALITY ASSESSMENT OF WORM GEAR DRIVES

Quality tests include:

- checking the geometry of the worm (Chapter 6);
- testing the meshing of the mated elements;
- checking the main operational characteristics of the drive.

8.1 TESTING THE MESHING OF THE MATED ELEMENTS

Following the checking of the geometry of elements, their meshing conditions were tested for the nominal centre distance $\underline{a}$. In the manufacturing precision of the mating elements, the deviations of the housing, especially of centre distance Δ_a and the perpendicularity Δ_γ of the axes, influence the shape and position of the contact area of mating teeth of the built-in couple. For this reason the investigation should be carried out in two stages, before and after build.

The meshing conditions of the mating elements before build were investigated on Maag and Klingelnberg testing devices. To adjust the contact area, the worm surface is smeared with Parisian blue and then, by braking the gear, it is rotated. On the flank of the worm gear teeth, the contact area becomes visible and can be investigated.

When the contact area proved to be adequate the elements were built within the housing and the second test started.

8.1.1 Building in the mating elements

After completion of the geometric investigation, the mating elements were built into their own housing. Satisfactory operation of components requires keeping to precise housing dimensions. The bearing supports on the shafts of the elements are shown in Figures 8.1 and 8.2.

During assembly or alternatively at change of bearings, to avoid damage occuring on the tooth flanks, no axial pressure can be applied on the teeth elements either during assembly or removal of the rolling bearings. To avoid these problems, the shafts in both cases (assembly or removal of bearings) should be axially supported. Another important consideration is to guarantee clean conditions when mounting the bearings.

The rim of the worm gear of high-power drives is typically manufactured from tin bronze. Care should be taken with the method of fixing the rim on the body of the worm gear, so it is advisable that the load pressures the rim against its supporting shoulder and never allows the connecting bolts to transfer the load between the rim and the body of worm gear.

As the housing is split at the symmetry plane of the worm (Figure 8.1), both below and above, it is suggested that self-aligning spherical roller bearings be included; also, care should be taken not to choose too small a distance between the bearing supports of the worm gear shaft in order to avoid or to reduce the side-swing of the worm gear resulting from contact forces on the mating teeth. The bearing support should be at clearance minimum in the axial direction in order to retain the adjusted contact area.

The housing should be designed to be rigid enough to guarantee proper meshing of the teeth, and to allow sufficient space for lubricating oil to penetrate (Dudley, 1984).

Removing bolts 11 and 12 can facilitate the oil change. The lower bearings are protected from solid contamination in the oil by oil-splash rings.

To adjust the contact area on the teeth, the lower bearing is fitted in a separate bearing housing 8 which, with the help of distance plates 9, can be adjusted and, with bolts 10, can be secured.

A proper bearing construction of the worm shaft requires the smallest possible distance should be taken between bearing supports to keep the bending deformation of the worm to the minimum possible value (Figure 8.2). This ensures that the contact conditions are near optimal.

The distance tube 4 can be used to adjust the axial clearance of the shaft, the axial size of it being determined on assembly.

8.1.2 Adjustment and position checking of contact area

Following assembly, the first task is to adjust for the correct contact area. For successful adjustment, the elements should be manufactured according to dimensions provided in the workshop drawing so that the tolerances are maintained. First, the axis of worm gear has to be positioned so that maximum clearance is left between mating teeth. Then the axis of the worm gear should be lowered by a small amount (this adjustment depends on the direction of rotation and on the value of lead). Having carried out these adjustments, the first checking of the contact area can commence.

Through the hole of the oil-level indicator (Figure 8.1) three teeth of the worm gear should be painted with Parisian blue (it is important to apply a paint of quality that forms a very thin layer but results in a good painting effect). To obtain a really thin layer of paint it is advisable to distribute the paint more or less equally over the flank surface using a piece of cloth. Then the worm gear, in the direction of operation, has to be rotated by a complete revolution by braking it slightly in the last quarter of the revolution so that the mated teeth press each other sufficiently. Having completed the revolution, the worm gear contact area can be seen through the checking hole. The position of the axis, starting in one direction in steps of 0.2 mm maximum, should be lowered or lifted. Experience shows that significant changes in the contact area results from a 0.1 mm change in the position of the axis.

The adjustment into the correct position of the contact area and the fixing of the worm gear axis is such an important task that the lower bearing is in-built and adjustable using a bolt in a separate bearing housing (Figure 8.1, item 8).

For each positional adjustment of the axis position it is always necessary to check the parallelism of the axial contact surface of the separate bearing housing the face surface of the complete worm gear housing. The checking of this is equally important during adjustment and fixing.

Having completed the contact area adjustment, the position of the worm gear should be fixed by selecting and fitting a set of plates of proper thickness (Figure 8.1, item 9).

During manufacture of the distance plates, it is necessary to ensure that the plates being used for a particular worm gear have been ground together while clamped. Having attained the correct contact area it is suggested that the remaining paint be removed from the worm gear and, by repeating the process, it is possible to recheck it. Rechecking is necessary also after assembling the final pack of distance plates and fixing the position of the axis.

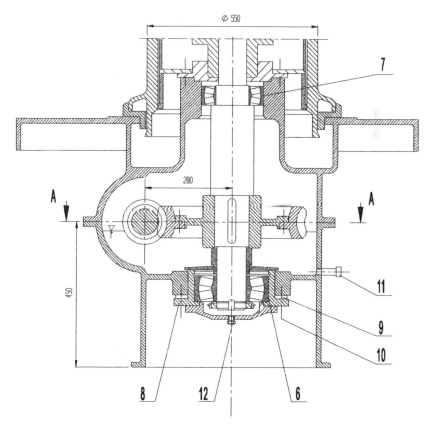

Figure 8.1 _Gearing support on worm gear shaft_
(Type DHG-550 wire drawing machine (DIGÉP) Hungary)

Because of normal manufacturing inaccuracies, the correct contact area cannot form on the contact surfaces and the probability of uniformity of worm gear drives manufactured in series is reduced. Apart from this and taking into consideration all the possible variants of inaccuracies, the contact area can be adjusted to be optimal. The contact area can be regarded as optimal if the axis of the worm gear can only be shifted a minimum distance in any direction, and the contact area will be worse than the adjusted optimal. Taking this into consideration, the general rule remains valid that the contact area should be shifted a little off-side to the worm gear teeth and it should extend to the largest possible area. The contact area of a newly manufactured unit is not normally a continuous area and it is rare for it to be extended to a really large area.

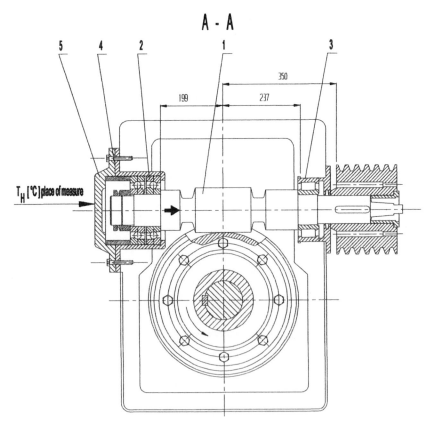

Figure 8.2 *Worm shaft as installed*
(Type DHG-550 wire drawing machine (DIGÉP) Hungary)

In the case where a contact area over a worm gear flank surface is a narrow trapezium or a triangular area, it is better that the higher face width of this area is situated towards the step as this improves the lubrication.

It should be mentioned that the opposite is disadvantageous with respect to lubrication. The different positions of the contact area, including the correct one, can be seen in Figure 8.3.

Regarding the symbols used in the figure, it should be noted that the arrows facing in the same direction represent the meshing direction of rotation. Two different worm gears mated with the same worm are symbolized by different direction of hatching lines. In the case of left-hand teeth, the adjustment of elements should plainly be carried out as a mirror view of that shown in Figure 8.3.

After adjustment that followed six to eight hours of idle running, several smaller areas begin to appear, eventually combining into one larger area, as the gear remains under load.

The final contact area, experience shows, appears after about 35 hours' running in. In the case of a wrongly adjusted contact area, a typical one is formed after a significantly longer period.

It should be noted that contact area of drives, working alternately in both directions, should be adjusted symmetrically at the midpoint.

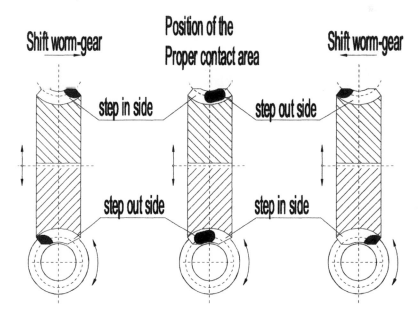

Figure 8.3 _Right-hand teething; adjustment of correct contact area_

Figure 8.4 *The contact area obtained on the tested, i = 4.8, worm gear drive before running in (The contact area is the black spot made by the black paint applied)*

The axial clearance of the worm gear shaft should be checked to keep the contact area in the previously adjusted position under load.

The worm shaft should have axial clearance because heat expansion during operation causes changes of size. For the same reason, preparation of clearance-free teeth is not recommended and the worm shaft may be fixed only at one end (Figure 8.2) in power-transmitting worm gear drives.

The most important and clearest precision-indicating characteristic of the power-transmitting worm gear drive is the contact area. The correct contact area is expressed as a percentage of the worm flank after running in, taking into consideration the manufacturing precision of the housing, because its tolerances highly influence the contact area formed.

The more precise the manufactured teeth, the bigger the contact area.

For the case of a localized contact area, apart from its extent, its position on the tooth flank is also important from the point of view of lubrication.

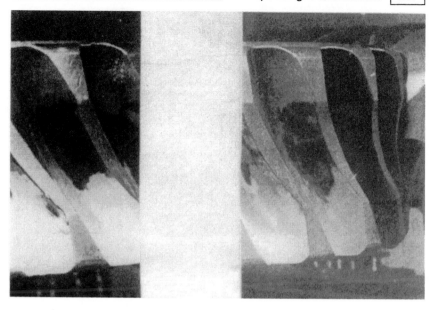

Figure 8.5 _The contact area obtained on the above tested worm gear drive after running in_

Figure 8.6 _The contact area obtained on the tested, i =11.67, worm gear drive before running in_

Figure 8.7 *The contact area obtained on the above tested worm gear drive after running in*

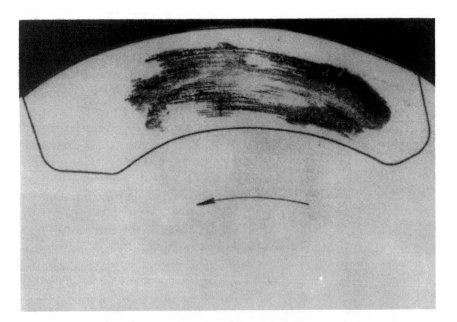

Figure 8.8 *The contact area obtained on the tested, i = 11.67, worm gear drive before running in (The photo was taken from the imprint of the image presented in Figure 8.6)*

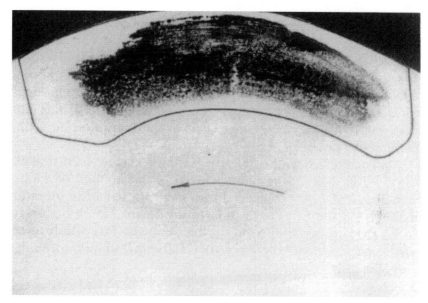

Figure 8.9 _The contact area obtained on the tested, i = 11.67, worm gear drive after running in (The photo was taken from the imprint of the image shown in Figure 8.7)_

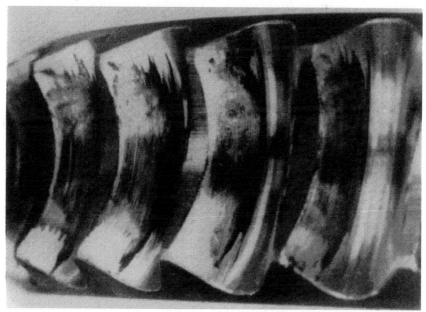

Figure 8.10a _The contact area on worm gear manufactured by fly-blade having thread-face surface (before running in) (m = 6 mm; a = 125 mm; z_1 = 1; H = 18.849 mm)_

Under nominal power transmission conditions, the contact area is shifted slightly towards the step-out side of the worm.

In the following figure, the contact areas obtained in tests are shown. Imprints were taken from contact areas making it possible to determine their value as a percentage, as summed up in Figure 8.10b(1). As a basis for comparison, in Figure 8.10b(2) the minimum value of the contact area cited in the literature (Krivenko, 1967) is also given.

Comparing the data in Figure 8.10b, it can be seen that the tested worm gear drives manufactured by the method developed here are satisfactory compared with the data from the literature or that from the CAVEX manufactured drives. The deviation for $i = 4.8$ drive originated from the fact that the face width of the worm gear tested was too large ($b_2 = 130$ mm), while for a similar drive it is only $b_2 = 90$ mm.

Measure of contact area obtained in test 1			
Type of worm	No starts on worm	Percentage of contact area	
	Z_1	Tooth length h (%)	Tooth length h (%)
Arched profile worm gear drive	$Z_1 = 5$ $i = 4.8$	48	90
	$Z_1 = 3$ $i = 11.67$	85.5	90

Minimum required contact area according to 2 (Krivenko, 1967)			
Type of worm	No starts on worm	Percentage of contact area	
	Z_1	Tooth length h (%)	Tooth length h (%)
In case of IT 7 accuracy	$Z_1 = 4$	45	60
	$Z_1 = 3$	50	60

Figure 8.10b *Contact areas during manufactured test drives and the minimum required contact area according to data cited in the literature*

8.2 CHECKING THE IMPORTANT OPERATIONAL CHARACTERISTICS OF WORM GEAR DRIVES

In this section we review the following operational tests:

- running in of the drive;
- determination of optimal level of oil;
- checking heat generation within the drives;
- determination of efficiency of the drives;
- determination of the noise level of the drives.

The measurements provided assessment of both worm gear drives with i = 4.8 and i = 11.67 on complete worm gear drives. To obtain comparable results, both drives were built into the same housing and the test conditions were kept constant.

For better comparison, the parameters were determined under the same conditions for the test drive with i = 11.67 for convolute worm, as well for i = 41 (spiroid worm) and for i = 39 (ZK2 worm) (Figure 8.26).

8.2.1 Running in of the drives

Having adjusted the contact area before loading the drive, it is necessary to run it in. It should be continually operated at idle running speed filled up to the required level with properly selected oil for a few hours. For running in, the prescription of the firm CAVEX was followed together with the results of previous experience gained, so the process as a whole was divided into four periods.

Period one, after eight hours' idle running, was followed by the second period of 10 hours with the operation under one-third of load. The third period lasted 15 hours under two-thirds load (in this case it is 26.4 kW) and the last period for another 15 hours under whole load (35 kW).

During the running in process, the following data were noted: shaft speed of the driving motor (n_m [s^{-1}]); shaft speed of the worm (n_1 [s^{-1}]); temperature of the house (T_H [°C]); and the ambient temperature (T_k [°C]).

During running in, small metal fragments are separated from the surface of the worm gear and settle into the oil. When creating the contact area during running in under load, significantly more heat is generated than during normal operation. During running at $\Delta T = 60$°C, 'over-warming' can be allowed. In practice it is advisable to keep a safety limit on temperature and ensure that it does not exceed $\Delta T = 70-80°C$ (temperature (T_H)). Under such circumstances it is necessary to start again with the oil temperature reduced to 50°C.

When the drive becomes heated beyond the higher than permitted temperature, a systematic check by measurement of all possible sources of overheating should be made.

Having completed running in, the temperature of the drained oil will be reduced to the operational temperature. After the oil is drained, both the bearings and the housing should be carefully washed with clean oil and replaced by oil of the same quality as has been used before.

In special cases it can happen that running in with full load occurs. Then, it is necessary to have about 20 hours' idle running after final build. During the first 50 hours under load, operations should be stopped to check the temperature and only continued after the system cools to 50°C.

For wire drawing, neither hour-to-hour stopping nor frequent checking is possible so wire drawing stages can only be transferred from the manufacturer to the place of use after running in under load has been carried out. Careful running in is crucial to the long service life of the drive. The time required for running in depends on the size of the drive and on operational conditions. In principle, running in should be continued until the contact area exhibits a joined-up image that reaches the prescribed percentage. Experience shows that running in increases the contact area and increases the efficiency.

The necessary checking by observation of the contact area of the worm gear can be done through the oil-level indicator window.

8.2.1.1 Review of the running in device

For the investigation of operational characteristics, the device shown in Figure 8.11 was designed and proved to be suitable for running in and could be used to determine efficiency of the drive.

The device shown in Figure 8.11 consists basically of the asynchronous electric motor 5 fixed on base frame 1 which drives the worm shaft 2, built in the drive investigated, by the variable Vee-belt drive (i_{el}). The direct current electric motor 4 plays the role of an electric generator, as a useful load. Between the output shaft (worm gear shaft) of the drive and the generator, a two-step accelerating Vee-

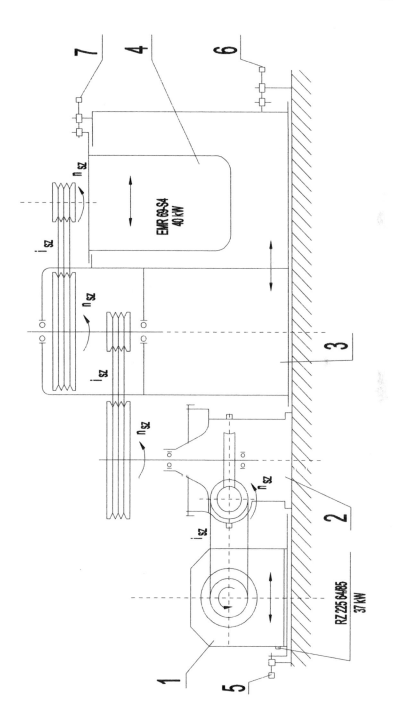

Figure 8.11 *Scheme for the running in and loading device*

belt drive is built in (i_{e2}, i_{e3}). By changing the value i_{e1}, different input shaft speeds for the worm shaft n_1 can be realized.

8.2.2 Determination of optimal oil level

According to the literature, for up to a peripheral worm speed of $v_1 = 10$ mpsec, it is suggested that oil bath lubrication is used. To choose a suitable oil viscosity, the tables found in the literature (Niemann, 1965) were employed.

To determine the idle running power consumption as a function of oil level, several measurements were carried out at constant operational temperature (at about 50°C) and at the most frequently used input shaft speed ($n_1 = 24.5$ rpsec). The literature provides recommendations on oil level for both lower and upper worm drives. The results of the measurements used to determine optimal oil level are presented in Figures 8.12 and 8.13. The optimal oil level for drive i = 4.8, according to Figure 5.1, falls into the range $\Delta l = 40$–50 mm, for drive i = 12, according to Figure 5.2, fits into the range $\Delta l = 35$–45 mm.

The decrease in idle running power P_0 with further reduction of the oil level is steep, but as lubricating conditions deteriorate, the noise level increases. It is important to keep the oil at a level to at least reach the dedendum cylinder of the worm. Excessive submergence increases the waste of energy, leads to oil turbulence and reduces the efficiency (Niemann, 1965).

8.2.3 Investigation of warming up of the drives

Only a part of the input power P_{in}, supplied through the worm shaft can be utilized as output power P_{out}, owing to the efficiency η of the drive. The difference is explained by the wasted energy P_w which is converted into heat. Its value is:

$$P_w = (1 - \eta) \cdot P_{in} \text{ (kW)} \qquad (8.1)$$

the drive having efficiency η for input power P_{in}.

A further reduction of the oil level generates a heat flow rate:

$$\dot{Q}_f = (1 - \eta) \cdot P_{in} \text{ (kW)} \qquad (8.2)$$

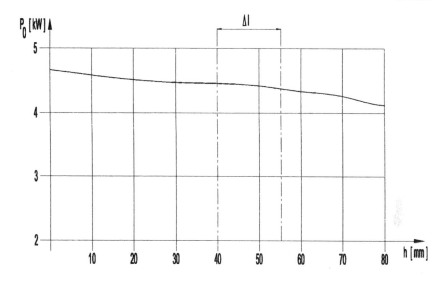

Figure 8.12 _Determination of the optimal oil level (Δl) for drive i = 4.8_

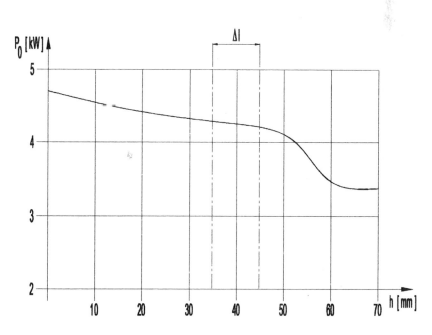

Figure 8.13 _Determination of the optimal oil level (Dl) for drive i = 11.67_

The heat flow rate generated increases the temperature of the drive T_H compared with the ambient temperature T_K. The part of the heat flow rate transferred into the ambient air can be expressed as:

$$\dot{Q}_e = A \cdot u \cdot (T_{HF} - T_k) \cdot 10^{-3} \quad \text{(kW)} \tag{8.3}$$

where:

T_{HF} is the allowable maximum operational temperature of the drive (°C),
T_K is the ambient temperature (°C),
u   is the coefficient of heat transfer of the house (W/m² °C),
A   is the surface area of the housing (m²).

It is not recommended to exceed the upper limit value of the housing temperature T_{HF}, because experience indicates that, as the viscosity of the lubricating oil with increasing temperature decreases steeply, the danger of seizing of the teeth is great. This is because lubricant protection decreases with decreasing viscosity.

For a temperature difference $\Delta T = (T_{HF} - T_K)$, the literature gives different values for ambient temperature $T_K = 20°C$. For a temperature difference Szeniczei (1961) suggests 40 to 50°C, Niemann (1965) mentions 60°C, while Krivenko (1967) thought 60–64°C an allowable value. But it is still not agreed at which point of the housing the temperature T_H should be interpreted.

To determine the load-carrying capacity, the ratio of the generated $(Q°_f)$ and conducted away $(Q°_e)$ heat flow rates should be investigated; this is an indicator by which to judge the efficiency of the drive. When the heat conducted away, $Q°_e$, is equal to heat generated, $Q°_f$, the construction is good. If the situation is the opposite, the arrangement is not a good one and a different approach to cooling is necessary to dispose of the heat excess $Q°_H = Q°_f - Q°_e$. There are several well-known solutions, like cooling by ventilation or to build cooling tubes into the oil bath .

For higher oil flow rate application, the use of an oil cooler both for submerged and oil-spraying lubrication can achieve increased heat conduction (and sometimes) a higher load-carrying capacity can also be achieved. The upper limit of load-carrying capacity is the power transmittable at the maximum operational temperature T_{HF}.

The maximum operational temperature T_{HF} according to different authors is: Szeniczei 70°C, Niemann 90°C, the CAVEX catalogue cites 90°C and Krivenko 90–100°C.

The value of T_H at different points of the housing can be different. According to safety measurements, the peak value was taken at the bearing housing (Figure 8.1, T_H measuring point).

As well as measuring T_H, the ambient temperature T_K was measured as a function of operational time at quarter, half and one-hour periods up to stabilization of the temperature of the house, ie when $T_{HF} = T_{stac}$, at the balanced heat state. During the tests, how the drives meet the requirements of designed, real operational conditions was investigated as well.

The values of T_H for given input shaft speeds n_1 and input powers P_1 were investigated to ascertain whether the drives can fulfil their task with the parameters used for design.

The input shaft speeds were $n_1 = 13.5, 24.5, 18.8$ rpsec, and the input power varied in the range $P_1 = 16...25$ kW. It was found that low input shaft speed was not critical from the point of view of heat generation so its detailed investigation was not necessary. At every critical shaft speed, the transferable power limit belonging to the temperature limit value T_{HF}, which is itself characteristic of the drive, can be determined. In Figure 8.14 the curve shown is obtained for measured temperatures T_H and T_K to reach the heat balance for the drive i = 11.67 at constant input shaft speed $n_1 = 24.5$ rpsec and at idle running power P_o.

Figure 8.14 shows that heat balance ($T_H = T_{stac}$) occurred at 51.5°C with ambient temperature of $T_K \approx 20$°C within about four hours of operation. The temperature of the housing T_H (and the oil temperature) increases with time and the heat balance can only be reached after several hours' operation, so the operation at heat-limit power can last only for a short period, or in the case of frequently interrupted operation, the time period is significantly longer than for continuous operation. The heating curve is characterized by the long-lasting temperature difference (ΔT). The results of other measurements are summarized in Figure 8.26.

8.2.4 Investigation of efficiency of drives

The last check on the power-transmitting worm gear drive is the determination of mechanical efficiency η_{wgd}. The measurements can be carried out on a test stand after running in the worm gear's own housing.

During these measurements, the experimental drive with i = 11.67, as well as the ruled surface (convolutes) drive with i = 11.67, were

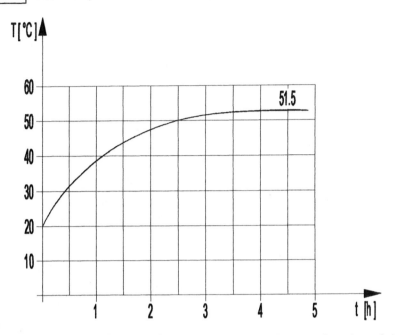

Figure 8.14 *The changes of housing temperature T_H as a function of time (t) at idle running*

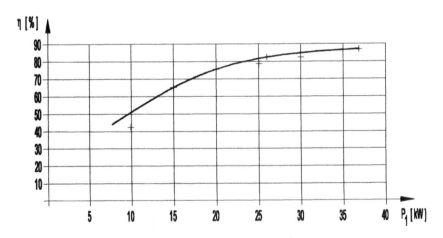

Figure 8.15 *Efficiency of the drive as a function of power P_1 supplied by the driving motor. The drive was built with arched profile in axial section of worm–worm gear. (Technical data: $i = 4.8$, $a = 280$ mm, $m = 16$ mm, $n_1 = 24.5$ rpsec; DHG-550 wire drawing stage, DIGÉP, Hungary)*

investigated and their efficiencies measured under equal conditions. To determine the efficiencies, the equipment shown in Figure 8.11 was used (Drobni, 1968). The values measured were the input and output electrical powers and the waste dissipated by individual units calculated.

The calculation method is as follows. The electrical power consumption of the driving machine was measured and the mechanical power is:

$$P_{mech} = \eta_m \cdot P_{electric\ motor} \tag{8.4}$$

The power supplied through the input shaft into the generator was calculated from the electric power measured and efficiency of the generator:

$$P_{g\ in} = \frac{P_{electric\ gen.}}{\eta_g} \tag{8.5}$$

The total efficiency of the unit is:

$$\eta_\Sigma = \frac{P_{g\ in}}{P_{mech}} \tag{8.6}$$

The total efficiency of the measuring system η_o depends on losses in the Vee belt drive between the shafts and motor and of the worm (ie η_{el}), the waste of the drive and its bearings (efficiencies η_{ec} and η_{e3}) and the waste of the bearings on the power-transmitting shaft ($\eta_{power\ transmitting\ shaft}$).

The catalogues of drive manufacturers contain the efficiency characteristics of the drive, including the built-in rolling bearings, so the losses of the bearings on the determination of the efficiency of the drive η_{wgd} will not be taken into account (except the bearings on the power-transmitting shaft) so the efficiency characteristics of the drive on the basis of measured and chosen data can be determined. Taking into consideration the waste, the total efficiency of the worm gear drive is:

$$\eta_{wgd} = \frac{P_{mech\ out\ wgd}}{P_{mech\ in\ wgd}} \tag{8.7}$$

where:

$$P_{mech\ out\ wgd} = \frac{P_{electric\ gen.}}{\eta_g \cdot \eta_{e2} \cdot \eta_{e3} \cdot \eta_{power\ transmitting\ shaft}} \tag{8.8}$$

$$P_{mech\ in\ wgd} = P_{electric} \cdot \eta_m \cdot \eta_{el} \tag{8.9}$$

After substitution we obtain:

$$\eta_{wgd} = \frac{P_{electric\ gen}}{P_{electric\ motor}} \cdot \frac{1}{\eta_g \cdot \eta_m \cdot \eta_{e1} \cdot \eta_{e2} \cdot \eta_{e3} \cdot \eta_{power\ transmitting\ shaft}} \tag{8.10}$$

The various efficiencies are:

η_m = 92.5% Asynchronous electric motors (based on catalogue data),
η_g = 90% Special AC engines (based on catalogue data),
η_e = 96% ten Bosch, M.: Maschinen Elemente,
$\eta_{power\ transmitting\ shaft}$ = 99.82%; this value was determined from bearing loads and shaft speeds and from calculated waste.

Substituting these values in equation (8.10) we obtain:

$$\eta_{wgd} = \frac{P_{electric\ gen}}{P_{electric\ motor}} \cdot \frac{1}{0.9 \cdot 0.925 \cdot 0.96 \cdot 0.96 \cdot 0.96 \cdot 0.9982} = \frac{P_{electric\ gen}}{P_{electric\ motor}} \cdot \frac{1}{0.735239} \cdot 100\% \tag{8.11}$$

Using the equation and knowing the measured values, the efficiency of the drive can be simply determined. For example, using the measurement data: at i = 4.8 gearing ratio, input shaft speed a_1 = 24.5 rpsec and P_1 = 37 kW input power with $P_{electric\ motor}$ = 23.72 kW measured for the electric generator, the following value of efficiency can be calculated using equation (8.11):

$$\eta_{wgd} = \frac{23.72\ kW}{37\ kW} \cdot \frac{1}{0.735239} \cdot 100\% = 87.2\% \tag{8.12}$$

The efficiency curve based on measurement is shown in Figure 8.15.

8.2.5 Investigation of noise level of drives

The need to reduce workplace noise has led, worldwide, to the introduction of safety codes related to permissible noise levels. Noise assessment is carried out with reference to the so-called *N curves*. These curves express the sound level in *decibels (dB)* as a function of frequency. In accordance with this criterion, continuous active noise measured in each octave range cannot exceed the ISO standard N-80 curve, which represents the value of the limit of hearing loss throughout an eight-hour shift.

The measurements can only be carried out under actual working conditions, so the measurements are usually disturbed by significant

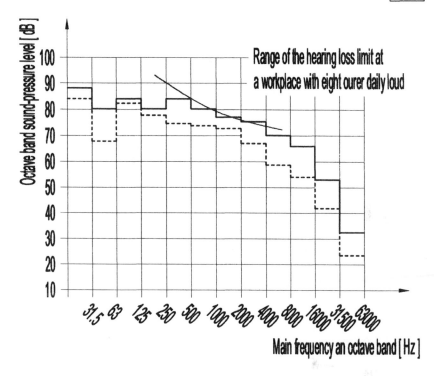

Figure 8.16 *Sound level, measured for different frequency bands, for operation of a circular profile in an axial section worm gear drive ($I = 11.67$, $a = 280$ mm, $n_1 = 24.5$ rpsec, $P_1 = 32$ kW)*

background noise. Therefore, for each frequency band, beside the common (resultant) noise level, the level of background noise should be measured too. The noise level of the investigated unit can be determined by knowing from measurement the two component noise levels. The measurements were carried out at a height of 1.5 m at a distance of 1m from the unit operated under idle running and with full load. The diagram in Figure 8.16 refers to measurements taken for i = 4.8, n_1 = 13.5 rpsec and P_1 = 35 kW.

In Figure 8.16, the continuous lines denote the sound level of the drive while the dashed lines are the background sound level. The diagram shows that the drives investigated, from the point of view of sound level, fulfil the necessary requirements.

Figure 8.17 *Dry wire drawing machine, Barcro system, driven by the worm gear drive designed and manufactured by DIGÉP*

Figure 8.18 *The Barcro system dry wire drawing machine with worm gear drive and additional gearbox*

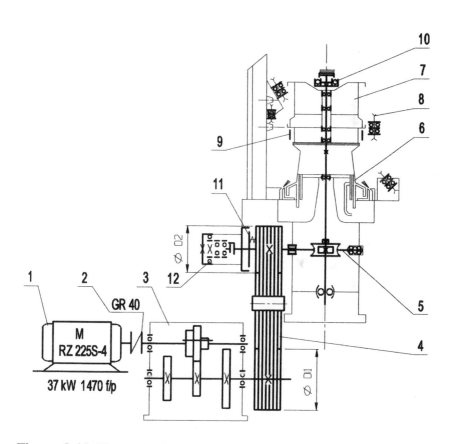

Figure 8.19 _Kinematical scheme of the Barcro system dry wire drawing machine driven by worm gear_

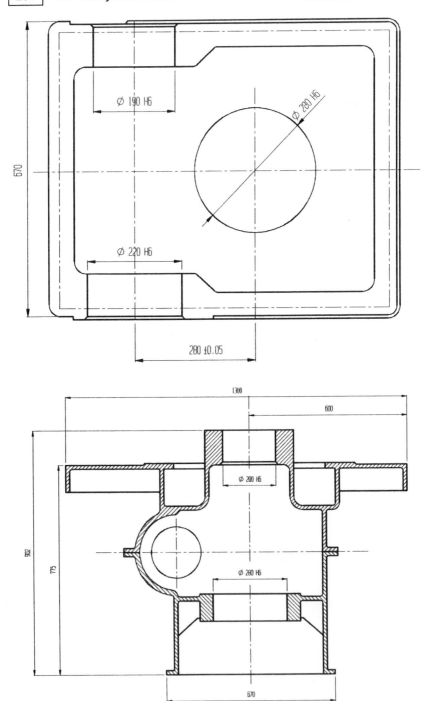

Figure 8.20 *Manufacturing tolerances of the drive housing*

Figure 8.21 _Checking the contact area of arched profile worm gear drives_ _(m = 9 mm; a = 280 mm; i = 25.5)_

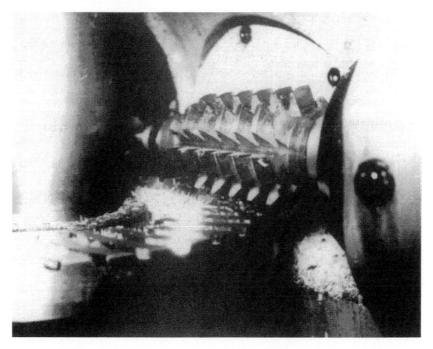

Figure 8.22 _Manufacturing of spiroid gear (KI type). Engineering material:_ _Bzö 12_

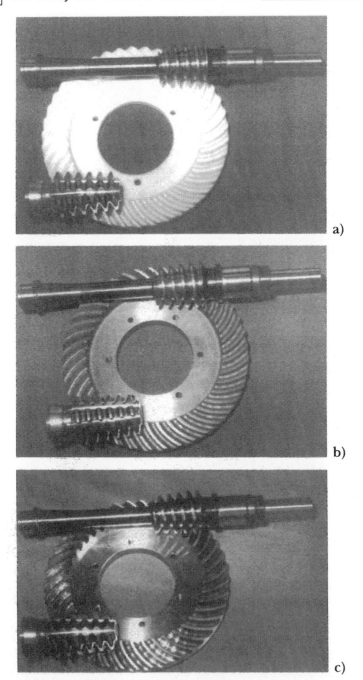

Figure 8.23 *The manufactured spiroid drive: a Archimedian (plastic wheel); b convolute (cast iron wheel); c involute (bronze wheel) and generating milling cutters*

Figure 8.24 _Investigation of the contact area of a spiroid drive on PSR-750 type equipment before running in_

Figure 8.25 _Investigation of the contact area of the drive after running in_

Type of worm gear drive	Gearing ratio z_2/Z	Module M (mm)	Diameter of the reference circle d_{01} (mm)	Shaft speed of the worm n_1 (s^{-1})	Characteristic measured idle running			Characteristic at load transmission			Contact area (%)			
					P_0 (kW)	T (°C)	Noise level	P_1 T (kW)	T (°C)	Noise level	Efficiency	Face width of teething b_2 (mm)	L_h (%)	h_k (%)
Worm gear drive with circular profile	24/5= =/4.8/	16	152	13.5	3.8	19	Adequate	26.6 / 35	34.4 / 43	Adequate	87.14	130	48	90
				24.5	4.6	26	Adequate	26.4 / 30 / 35	55 / 56 / 54	Adequate	86.44			
in axial selection a=180 mm	/11.67/	12.5	97.5	13.5	-	-	Adequate	19	41	Adequate	82.65	90	85.5	90
				24.5	4.11	33	Adequate	16 / 28 / 32	51.5 / 58.5 / 61.5	Adequate				
Convolute worm gear drive a=280 Mm	35/3	13	104	24.25	4.2	25.3	Adequate	22	36	Adequate	73.52	90	-	-
Spiroid*	41	5	58.5	20.83	-	-	Adequate	760	60	Adequate	80/70 E/H**	40	90	90
ZK2 Type*	39	4	56	20.83	-	-	Adequate	550	-	Adequate	68	35	45	45

*Data from literature (Hegyháti; 1985, 1988).
**E = concave–convex surfaces in contact;
H = convex–convex surfaces in contact.

Figure 8.26 Operational characteristics of drives designed and manufactured by us

9

SUMMARY OF RESULTS
OF RESEARCH WORK

In summarizing the analytical and numerical results of the theoretical investigations and the experience drawn from the manufacturing and assessment processes, one can come to the following conclusions:

1. Applying the newly developed method, the grinding wheel profile is appropriate to back generation from the worm-generated profile. This guarantees independence from wheel truing (diameter changes) that in equivalent helicoid surfaces results from wrapping. This method eliminates the faults and difficulties that arise in previously used processes. This type of grinding wheel profiling was first used in the machining technology of arched profile worm gear drives.

   Independence from the effect of wheel diameter changes as is present in all equivalent helicoid surfaces is obtained with the worm as a result. The equations of the contact curve of the grinding wheel and the worm during machining and the surface of the grinding wheel have been determined.

2. During grinding of the worm gear tool, the author used the back generation from worm method as well, so the worm and worm gear formed conjugated surfaces. The equations of the other element of the mating pair, the worm gear generating tool (face plane, back surface, cutting edge), were determined both for fly cutter tool with a plane faced surface and the thread face surface tool which, in practice, is a generator worm gear milling cutter.

   This tool, necessary for machining, was made in the way described in this book and the results obtained testified to its effectiveness.

In the following cases, both the tools and the elements were manufactured:

a = 125 mm,	m = 7.5 mm,	$z_1 = 1$
a = 280 mm,	m = 12.5 mm,	$z_1 = 3$
a = 80 mm,	m = 16 mm,	$z_1 = 5$
a = 280 mm,	m = 9 mm,	$z_1 = 2$

Twenty worm gears were investigated for the effects of distortion caused by re-sharpening using radial grinding-backward of the tool cutting edge. Our experiences show that deviations due to removing the total re-sharpening reserve remain within the allowable tolerance. So the geometric shape of the tool can be regarded as that required.

The author designed a CNC-controlled grinding wheel-generating device that made it possible to generate an helicoidal surface of arbitrary profile. The essence of operation of this universal device was to generate an appropriate wheel profile. A computer-aided design workplace was created that was capable of producing a two-variable function, the tool surface, using proper software, starting from the generator curve of the workpiece, the lead on it and the number of starts (the direct method) as well as for generation of the input data necessary for tool manufacture on CNC wheel truing equipment. In this way CAD/CAM integration of the design and manufacture was created and a CAQ phase could be combined with it.

The design workplace in the CAD phase is able to solve problems when given the surface or generator curve of the generating tool for the worm to be designed and, by applying the principle of double meshing, the surface of the workpiece as a two-variable function has to be created (indirect method).

At the least the CNC grinding wheel generating device is the key element of the CAD/CAM/CAQ integration which creates the future possibility of a grinding manufacturing cell to combine into a CIM system. The theoretical structure of this scheme is shown in Chapter 6. Patenting of this CNC device has been carried out.

3. The mechanically operated device developed for wheel truing can be applied with advantage in mass production too. Using the device the cutting edge radius (S_{ax} and K parameters) can be manufactured within the usual range of mechanical engineering.

No special devices are required for checking geometric sizes of worms, because employing conventional measuring devices (eg Klingelnberg) the measurement in axial section can be carried out. The precision of geometric sizes of the designed and manufactured driving elements (worm and worm gear) prove that this machining technology is suitable for use in industry. The shape deviations of manufactured components proved to be a fraction of the allowable tolerance (Dudás, 1988a).

4. Results of measurement indicate that manufactured driving elements compared to other type drives of similar size, regarding efficiency and power transmitted at heat limit, are better than the others. For i = 4.8, the drive characteristics are similar to drive i = 5.8 while for i = 11.67 drive, it is better.

5. The efficiency of the manufactured drives is significantly better than the drive with ruled convolute surface. This has an important bearing on energy conservation.

6. The parameters of the contact area, for the drives investigated, are better than those quoted in the literature, so their size is satisfactory. At the same time, their position shows that the contact area is localized, and the creation of the pressure in the lubricating material is advantageous from the point of view of efficiency. The contact area of this drive is satisfactory.

7. The shape and position of the contact area, improved as a result of running in under gradually increased load for an appropriate time before taking them into use, improved as required.

8. According to the diagrams, sound level of the drives remains below the value of the N-80 hearing loss limit, so they meet the requirements of environment protection.

9. The drives manufactured and tested according to the method described in this book have advantageous operational characteristics as are summarized in Figure 8.26.

10. The various drives manufactured for the Metallurgical Works at Salgótarján in 1974, which are still working, were built into the drive system of dry wire drawing stands. The DIGÉP company manufactures these types of drives and they were designed and manufactured with worm drives with arched profiles of different scale for different factories. For example, the drives of pulverisers manufactured by the milling machine manufacturers in Jászberén, the drives of metal plate blanking machines at the light metal works at Székesfehérvár, the drives of conveyors of the Agricultural Machines Development

Centre at Monor. All of these drives operate with good operational characteristics.

11. Both the manufacturing and the mechanically operated wheel profiling equipment from Hungary have been patented in several countries.

12. The manufactured drives and tools have been exhibited in 1978 at the BNV (Budapest International Fair), in 1985 at the SZASZ, in 1988 in the exhibition at the Heavy Industrial University of Miskolc (Faculty of Mechanical Engineering), as well as in June 1988 at Graz at the scientific exhibition.

13. The method for measurement and testing of geometric sizes of helicoidal surfaces is suitable for application in any firm equipped with a three-dimensional (3D) measuring machine. On the basis of the general mathematical model (Chapter 4) it has become possible to check helicoids on three-dimensional measuring machines. The test confirmed that measuring machines designed and manufactured for universal use equipped with proper software can be adapted for correct assessment of helicoidal surfaces without special, additional equipment (Chapters 6 and 8).

14. It is suggested, from the results obtained, that in 'clearance-less' drives (by simple axial displacement of the worm) spiroid drives can be employed.

15. The machining geometric characteristics of conical (spiroid) drives were investigated in detail using the so-called kinematic method to detect the possibility of manufacturing geometrically exact surfaces using continuous grinding wheel truing. Mathematical equations were derived that make it possible to calculate the position of contact curves during machining and to determine characteristic parameters also. The software developed, besides facilitating the calculations of the machining geometry, assists determination of the quasi-optimal technological parameter in an interactive way.

16. The mathematical model developed from the point of view of general validity and its general capacity has the following characteristics:

   - it can be used to investigate the manufacturing geometry of the elements of both cylindrical and conical worm gear drives independent of whether their surface is a ruled or non-ruled one;

- it is suitable for the analysis of machining geometry of cylindrical and conical tools;

- using the model, both cylindrical and conical helicoidal surfaces of arbitrary profile can be treated using a common mathematical tool.

Finally, and most important, the relative positions and displacement of the coordinate systems, with correct choice of parameters, are capable of modelling different worm gear drives independent of their profile or basic form (cylindrical, conical, globoid or helicon).

When modelling them, the contact curves can be determined and the mesh can be evaluated. So by changing the parameters it is possible to determine the optimal operational and meshing conditions.

REFERENCES

Altmann, F G (1937) Bestimmung des zahnflankeneingriffs bei Allgemeinen schraubengetrieben, *VDI Forschung aus dem Gebeit des Ingenieurwesens*, No 5

Abadziew,W and Minkow, K (1981) O geometrii wintowüh powerhnoste spiroidnüh peredac, *Teoreticna i prilozna mechanika*, No 2, Bulgarska Akademia, Sofia

Bakondi, K (1974) *Design of Relieved Hobs and Tooth Cutting Tools* (in Hungarian), Tankönyvkiadó, Budapest

Bálint, L (1961) *Planning of Machining* (in Hungarian), M Szaki Könyvkiadó, Budapest

Bányai, K (1977) Machining geometry and checking of cylindrical worms (in Hungarian), university doctoral dissertation, Miskolc

Bär, G (1977) Curvatures of the enveloped helicoid, *Mechanism and Machine Theory*, **32** (1), pp 111–120

Bär, G (1996) *Geometrie-Eine Einführung in die Analytische und Konstruktive Geometrie*, B.G. Teubner Verlagsgesellschaft, Leipzig, Stuttgart

Bercsey, T (1977) Theory of toroid drives (in Hungarian), candidate dissertation, Budapest

Bercsey, T and Horák, P (1999) Error analysis of worm gear pairs, 4th World Congress on Gearing and Power Transmission, CNIT, Paris, 16–18 March

Bilz, R (1976) Ein beitrag zur entwicklung des globoid-schneckengetriebes zu einem leistungsfähigen element der modernen antriebstechnik, Diss.B, TU Dresden

Bluzat, J P (1986) Rectification des surfaces heliocoidales d'une vis profilage par meule annulaire, 2 eme Congres Mondial des Engrenages, Paris, 1, pp 719–32

Boecker, E and Rochel, G (1964) Messprobleme bei der Fertigung von schnecken-getrieben, _Werkstatt und Betrieb_, No 2, pp 153–6

Boecker, E and Rochel, G (1965) Werkzeugfragen bei der fertigung von schnecken-getrieben, _Werkstatt und Betrieb_, No 1, pp 11–13

Bohle, F (1956) Spiroid Gears and Their Characteristics Machinery, 6 January

Bohle, F and Saari, O (1955) _Spiroid Gears – A New Development in Gearing_, AGMA Paper 389.01

Breitung, W (1993) Bedeutung offener CAD-Systeme für die Integration der Rechnerunterstützung von Konstruktion und Fertigung, CAD-CAM 10, Sonderteil in Hanser-Fachzeitschriften, pp 6–10

Buckingham, E (1949) _Analytical Mechanics of Gears_, McGraw-Hill, New York/Toronto/London.

Buckingham, E (1960) _Design of Worm and Spiral Gears_, The Industrial Press, New York

Buckingham, E (1963) _Analytical Mechanics of Gears_, 2nd edn, Dover Publications, New York

Capelle, J (1949) _Theorie et Calcul des Engrenages Hypoids_, Edition Dunod, Paris, 1/74

Crain, R (1907) Schraubenräder mit geradlinigen eingriffsflächen, _Werkstattstechnik_, Bd 1

Csibi, V I (1990) Contribution to numerical generation of helical gearing with any profiles (in Romanian), Phd dissertation, Technical University of Cluj-Napoca

Distelli, M (1904) Über instantane schraubengeschwindigkeiten und die verzahnung der hyperboloidräder, _Zeitschrift Math und Phys_, **51**

Distelli, M (1908) Über einige sätze der kinematicshen geometrie, welche der verzahnungslehre zylindrischer und konischer räder zugrunde liegen, _Zeitschrift Math und Phys_, **56**

Drahos, I (1966) _Construction of mating surface and contact curve set of Litvin type worm drive_ (in Hungarian), reprint of Hungarian Language Publication of NME, **XII**

Drahos, I (1958) Basis of geometrical dimensioning of hypoid bevel gear pairs (in Hungarian), university doctoral dissertation, Miskolc

Drahos, I (1967) Die grenze des eingriffsbereiches an der schneckenzahnfläche zylindrischer schneckentriebe, _NME Idegennyelvü Közleményei_, **27**, Miskolc

Drahos, I (1973) Ermittlung der eingriffsfläche und der berührlinienschar des TU-ME globoidschneckengetriebes wissenschaftliche, _Zeitschrift der TU Dresden_, **22** (2), pp 325–36

Drahos, I (1981) Eine systematik der verzahnungen mit sich kreuzenden achsen, vom standpunkt der kinemtischen geometrie aus betrachtet wiss, *Zeitschrift der TU Dresden,* **30** (4), 97–103

Drahos, I (1987) Basis of kinematical machining geometry (in Hungarian), academical doctoral dissertation, Miskolc

Drobni, J (1967) Globoid worm drives which can be grinded (in Hungarian), candidate dissertation, Miskolc

Drobni, J (1968) Calculation of cylindrical worm drives having arched profile (in Hungarian), *Publication of Department of Machine Elements of NME,* No 194, Miskolc

Drobni, J and Szarka Z (1969) Formation of restricted tooth contact surface in case of different worm drives (in Hungarian), 2nd Conference on Gears, Budapest

Dudás, I (1973) Simplified production and qualification of worm drive having arched profile (in Hungarian), university doctoral dissertation, Miskolc

Dudás, I (1976) Proper running in of power transmitting worm drives (in Hungarian), *Technical Publications of DIGEP,* **1**, pp 17–25

Dudás, I (1980) Development in tooling and in machining of worm drives having arched profiles (in Hungarian), candidate dissertation, Miskolc, pp 153 + 30

Dudás, I (1986a) Die Analyse der Werkeug- und Fertigungs-geometrie von Spiroidgetrieben. 7. Vortragstagung mit internationaler Beteiligung Fertigung und Gütesicherung im Zahnradgetriebebau, Magdeburg, 24–5 September, pp 215–21

Dudás, I (1986b) Analysis of tool geometry and machining geometry of spiroid drives (in Hungarian), *Gépgyártástechnológia,* **26**, pp 166–9

Dudás, I (1986c) Manufacturing and analification of drives with good efficiency and high load capacity, Department of Production Engineering at Technical University for Heavy Industry, 16–18 June, pp 155–67; 1st International Conference on Reliability and Durability of Machines and Machinery Systems in Mining, Szozyrk, Seria, Görnictwo.z, Poland, 16–18 June, p 143

Dudás, I (1986d) Questions of production geometry of spiroid drives, Correferatum prepared for the assembly of sub-committee of Power Transmissions of Machine Design Committee of Technical Sciences, Department of MTA Budapest (in Hungarian), 29 May

Dudás, I (1986c) Geometrisch richtige Herstellung von Schraubenflächen. IX. Nauhowa Szkola Obrobki Sciernej Krakow Wrzesien, 16–20 September, pp 81–6

Dudás, I (1988a) CNC grinding wheel dressing equipment, procedure for continuous and non-continuous control when grinding (in Hungarian), Patent, Number in Hungary 207 963, 21 September

Dudás, I (1988b) Theory of production of helicoid surfaces (in Hungarian), academical doctoral dissertation, Miskolc, p 250

Dudás, I (1988c) Design and manufacturing of helicoid surfaces and their tools using a CAD/CAM system, International Conference on Engineering Design, ICED '88, Budapest, 23–5 August, p 8

Dudás, I (1988d) Herstellung von helicoidflächen und werkzeugen unter nutzung eines CAD/CAM systems. III. Intersymposium 'Untersuchung von Werkzeugen '88', Krakow-Muszyna, 13–15 October, pp 32–6

Dudás, I (1988e) Korrekte abwälzfräser zur korrekten schnecken, Nass '88, Werkzeugkonferenz, Koszalin, 26–8 May

Dudás, I (1991) Manufacturing of Helicoid Surfaces in CAD/CAM System, International Conference on Motion and Power Transmission, MPT '91, Hiroshima, Japan, 23–6 November, pp 339–44

Dudás, I and Ankli, J (1978) Development of grinding wheel dressing equipment for worm drives having arched profile, accepted and applied innovation (in Hungarian), DIGEP A-2843, Miskolc

Dudás, I and Bányai, K (1994) Manufacturing of helical surfaces in flexible production system, ICARV '94, Singapore, 8–11 November, pp 1036–8

Dudás, I, Berta, M and Cser, I (1993) The condition and development problems of industrial process CAD-system, _Rezanie i Insztrument 47/93_, Harkov, ISSN 0370-808X, 9.88-94

Dudás, I, Cser, I and Berta, M (1998) Production of rotational parts in small-series and computer-aided planning of its production engineering, _Manufacturing_, Boston, Massachusetts, 1–5 November; ISSN 0277-786X; ISBN 0-8194-2979-1; SPI – the International Society for Optical Engineering, pp 172–7

Dudás, I and Csermely, T (1993) Qualification of CNC grinding wheel dressing equipment (in Hungarian), _Production Engineering_, **05–06**, pp 219–23

Dudás, I, Csermely, T and Varga, Gy (1994) *Computer Aided Measurement Techniques of Experiments Performed by Tools with Defined Geometry*, Computational Mechanics Publications, Southampton, Boston, ISBN 1 853123676, ISBN 1 562522914, Mechatronics Conference, Budapest, 21–3 September, pp 35–41

Dudás, I, Drobni, J, Ankli, J and Garamvölgyi, T (1983) Equipment and procedure for profiling of grinding wheels used for production of geometrically proper worm driving pairs having arched profile in the axial section (in Hungarian), Service Patent, No 170 118, date of announcement: 27 December

Dudás, L (1991) Realuson of geometrical problems of contacting surfaces using the reaching model, candidate dissertation, Miskolc, p 144

Dudley, D W (1954) *Practical Gear Design*, McGraw-Hill, New York

Dudley, D W (1961) *Zahnräder-Berechnung, Entwurf und Herstellung nach Amerikanischen Erfahrungen*, Springer-Verlag, Berlin

Dudley, D W (1962) *Gear Handbook: The Design, Manufacture, and Application of Gears*, McGraw-Hill, New York

Dudley, D W (1984) *Handbook of Practical Gear Design*, McGraw-Hill, New York

Dudley, D W (1991) *Dudley's Gear Handbook*, ed. D P Townsend, McGraw-Hill, New York

Dudley, D W and Poritsky, H (1943) On cutting and hobbing gears and worms, *Journal of Applied Mechanics*, Nos 3 and 4

Dyson, A (1969) *A General Theory of the Kinematics and Geometry of Gears in Three Dimensions*, Clarendon Press, Oxford

Erney, Gy (1983) Gears (in Hungarian), M Szaki Könyvkiadó, Budapest, p 460

Flender-Bocholt (1969) Getriebe und Antriebselemente V.239.D7.69. A. Frioedr Flender and Co Bocholt

Gansin, W A (1970) Linii kontakta evolventnoj spiroidnoj peredaci, *Mechanika Maschin*, No 9, p 127–32

Gansin, W A (1972) Sintezu evolventnoj spiroidnoj peredaci, *Mechanika Maschin*, pp 31–2

Georgiew, A K and Goldfarb W I (1972) Aspektü geometriceskoj teorii i resoltatü isledo-vanijaspiroidnüh peredac s zylindriceskimi cervjakami, *Mechanika Maschin*, Moszkva, No 00, pp 31–2

Georgiew, A K and Goldfarb, W I (1974) Kisledovaniju ortogonalnoj spiroidnoj peredaci s cylyndriceskim cervjakom, imejusim witki idealno-peremennowo saga, *Mechanika Maschin*, Moszkva, No 45

Gochman, H I (1886) _Theory of Gearing Generalised and Developed Analytically_ (in Russian), Odessa

Goldfarb, V I and Spiridonov, V M (1996) Design of the two-stage spiroid gear units and gear motors, Proceedings of International Conference on Gears, _VDI Berichte_, No 1230, pp 579–86

Golubkow, N S (1959) Nekotorüje woprosü geometrii zaceplenija cerwacno-spiroid-nüh peredac, _Izw. Wuz. Masinostroenije_, No 8

Grüss, G (1951) Zur kinetik des rollgleitens, _Zeitschrift für Angew Math. Mech_, **31**

Hegyháti, J (1985) Analysis of meshing conditions of spiroid driving pairs (in Hungarian), 6th Symposium on Tools and Tool Materials, Miskolc, pp 114–23

Hegyháti, J (1988) Untersuchungen zur Anwendung von Spiroidgetrieben, Diss. A.TU, Dresden, p 121

Höhn, B R, Neupert, K and Steingröver, K (1996) Wear load capacity and efficiency worm gears, _VDI Berichte_, No 1230, pp 409–26

Hoschek, J (1965) Zur ermittlung von hüllflächen in der räumlichen kinematik, Monh. für Mathematik, **69**

Jauch, G (1960) Meridiankonstruktion rotierender werkzeuge zur herstellung von schraubenflächen, Österreichische Ing. Archiv., **14**

Juchem, H (1987) Düsseldorf: Einsatzverhaltenkeramisch gebundener CBN-schleifscheiben, _Werkstatt und Betrieb_, **120** (1), pp 85–8

Kacalak, W and Lewkowitz, R (1984) Profilmodifikation geschliffener gewindeschecken, _Werkstatt und Betrieb_, **117** (2), pp 85–8

Kienzle, O (1956) Die grundpfeiler der fertigungstechnik, _Werkstattstechnik und Maschinenbau_, **46**

Kolchin, N I (1949) Analytical investigation of planar and spatial gearing (in Russian), _Mash-giz_

Kolchin, N I (1968) _Nekotorüe voproszü geometrii, kinematiki, raszcseta i proizvodsztva_, Leningrad, p 362

Kornberger, Z (1962) Preklanie slimakowe konstrukcije wykovanie Warsawa, _Wyd-Nauk Techn_, **340**, pp 115–29

Krivenko, I Sz (1967) Novüe tipü cservjacsnüh peredacs na szudah, _Izd. Szudoszrovenie_, Leningrad

Kyusojin, A, Munekata, K, Tanaka, M and Toyama, A (1986) A finishing method of rotor in screw compressor by fly tool, _Bulletin of JSME_, **29** (253), pp 2301–5

Lange, S (1967) Untersuchung von Helicon- und Spiroidgetrieben mit Abwickelbaren Schneckenflanken (Evolventtenschnecken)

nach der Hertzschen und der Hydrodynamischen, Theorie Diss., TH Munchen

Leroy, B, Boyer, A, Caracci, G and Depres, D (1986) Optimisation et CAO des engrenages spiroides a axes orthogonaux ou non, 2 eme Congres Mondial des Engrenages, Paris, **2**, pp 759–68

Lévai, I (1965) Gears – produced by hob having ruled surface – realizing movement transmissions between skew axes (in Hungarian), candidate dissertation, Miskolc, p 83

Lévai, I (1980) Kinematical theory of meshing of teeth and its application for the design of hypoid drives (in Hungarian), academical doctoral dissertation, Miskolc, 1/153

Lierath, F and Popke, H (1985) Technologische Untersuchungen und Konstruktive Lösungen zum Einsatz des Hartmetalls beim Wälzfräsen. VI. Werkzeug und Schneidstoffe Symposium, Miskolc, 27–29 August, p 61

Litvin, F L (1968) *Theory of Gearing* (in Russian), 2nd edn, Nauka, Moscow

Litvin, F L (1972) *Theory of Gear Mating*, M Szaki Könyvkiadó, Budapest

Litvin, F L (1994) *Gear Geometry and Applied Theory*, Prentice Hall, Englewood Cliffs, NJ

Litvin, F L and Feng, P-H (1996) Computerised design and generation of cycloidal gearing, *Mechanism and Machine Theory*, **31** (7), pp 891–991

Litvin, F L and Guo, K (1962) Investigation of meshing of bevel gear with tared teeth (in Russian), *Theory of Machines and Mechanisms*, **92** and **93**

Lotze, W, Rauth, H H and Ertl, F (1996) New ways and equipments for economic 3D gear inspection, *VDI Berichte*, Nr 1230, pp 1021–30

Magyar, J (1960) Meshing of elements having helicoid surfaces (in Hungarian), candidate dissertation, Budapest

Maros, D, Killmann, V and Rohonyi, V (1970) *Worm Drives* (in Hungarian), M Szaki Könyvkiadó, Budapest

Müller, H R (1955) Zur kinematik des rollgleitens, *Archiv. Mathematik*, **6**

Munro, R G (1991) Single analysis of some of Niemann's noise measurements of spur gears, International Conference on Motion and Power Transmissions, Hiroshima, Japan, pp 10–14

Nelson, W D (1961) *Spiroid Gearing Machine Design*, 1 February, 1, 16 March

Niemann, G (1956) _Untersuchung von Zylinderschneckentriebe mit Rechtwinklig sich Kreuzenden Achsen_, Braunschweig, p 153

Niemann, G (1965) _Maschinenelemente Band 2. Getriebe._ Springer-Verlag, Berlin, Heidelberg, New York

Niemann, G and Heyer, E (1953) Untersuchungen an schneckengetrieben, _VDI_, No 6, pp 147–57

Niemann, G and Winter, H (1983) _Maschienenelemente Band 3_, Springer-Verlag, Berlin, Heidelberg, New York, Tokyo

Niemann, G and Weber, G (1942) Schneckengetriebe mit flüssiger reibung, _VDI Forschungsheft_, **412**

Niemann, G and Weber, G (1954) _Profilbeziehungen bei der Herstellung von Zylindrischen Schnecken, Schneckenfräsern und Gewinden_, F. Vieweg, Braunschweig

Olivier, Th (1842) _Theorie Geometrique des Engrenager_, Paris

Ortleb, R (1971) Zur verzahnungs- und fertigungsgeometrie Allgemeiner zylinderschneckengetriebe, dissertation, TU Dresden

Patentschrift, Deutsches Patentamt, No 905444 47h 3

Patentschrift, Deutsches Patentamt, No 855527, 27h

Perepelica, B A (1981) Otobrazsenija affinnogo prosztransztva v teorii formoobrazovanija poverhnosztej rezaniem, _Harkov Vusa Skola_

Predki, W (1986) _Berechnung von Schneckenflankengeometrien Technical_, 22, pp 28–32

Reuleaux, F (1982) _Der Konstrukteur_, Vieweg Sohn, Braunschweig

Russian Patent No 139.531

Saari, O E (1956) Mathematical background of spiroid gears, _Ind. Math Series_, Detroit, Michigan

Schwägerl, D (1967) Untersuchung von Helikon und Spiroidgetrieben mit Trapezförmigen Schneckenprofil nach Herzschen und der Hydrodinamischen Theorie Duss, TH Munchen

Schwaighofer, R P, Kalin, A, Ledermann, P and Metzger, J L (0000) Polykristalline diamant- und bornitrid-werkzeuge, _Präzisions-Fertigungstechnik aus der Schweiz_, pp 79–84

Simon, V (1996) Tooth contact analysis of mismatched hypoid gears, Proceedings of the 7th International Power Transmission and Gearing Conference, San Diego, California, 6–9 October, pp 789–98

Siposs, I (1977) Toothing of the elements of globoid worm drives on CNC generating milling machine (in Hungarian), university doctoral dissertation, Miskolc

Stübler, E (1911) Geometrische probleme bei der verwendung von schraubenflächen in der technik, *Z. Math. und Phys.*, Band 60

Stübler, E (1922) Über hyperboloidische verzahnung, ZAMM 2

Su, D and Dudás, I (1996) Development of an intelligent integrated system approach for design and manufacture of worm gears, Proceedings 9th International Conference on Tools, Miskolc, Hungary, 3–5 September

Szeniczei, L (1957) *Power Transmission Worm Gear Drives* (in Hungarian), M Szaki Könyvkiadó, Budapest

Szeniczei, L (1961) *Conjugated Tooth Profiles* (in Hungarian), Hungarian Language Publications of NME, V, Miskolc

Tajnaföi, J (1965) Principles of movement generating features of machine tools and some applications (in Hungarian), candidate dissertation, Miskolc

Umezawa, K, Houjoh, H, Ichikawa, N and Matsumura, S (1991) Simulation of rotational vibration of a helical gear pair transmitting light loads, Proceedings 3rd JSME MPT International Conference, Hiroshima, pp 85–91

Varga, I (1961) *Some Suggestions for the Theory of Cylindrical Worm Gearing having Ruled Surfaces* (in Hungarian), Hungarian Language Publications of the NME, V, pp 371–89

Weinhold, H (1963) Zur fertigungsgeometrischen deutung technologischer prozesse, *Fertigungstechnik und Betrieb*, No 3

Wildhaber, E (1926) Helical Gearing, US Patent No 1,601,750

Wildhaber, E (1948) *Osnovü Oaxoplonija Konicesküh i Gipoidnüh Peredac*, Masgiz, Moskau

Wildhaber, E and Steward, L (1926) The design and manufacture of hypoid gears, *American Machinist*, **64**

Willis, R (1841) *Principles of Mechanism*, Cambridge, London

Winter, H and Wilkesmann, H (1975) Berechnung von schneckengetrieben mit unterschiedlichen zahnprofilen, *VDI-ztschr*, **117** (10)

Wittig, K H (1966) Zur geometrie der zylinderschnecken, *Maschinenmarkt*, **72**

Zalgaller, V A (1975) *Teorija Ogibajusih*, Nauka, Moskau

FURTHER READING

Bakondi, K. (1981) Development of production procedures and constructional forming of broaching tools (in Hungarian), academical doctoral dissertation, Budapest

Colbourne, J R (1974) The geometry of trochoid envelopes and their application in rotary pumps, *Mechanism and Machine Theory*, **9** (3 and 4), pp 421–35

Dudás, I (1974) *Grinding of Worm having Arched Profile* (in Hungarian), Publication of 3rd Conference on Gears, Budapest, pp 61–5

Dudás, I (1978) Geometric problems of tool for machining worm having arched profile (in Hungarian), Proceedings of the 4th Colloquium on Tools and Tool Materials, Miskolc, 9 January–31 August, p 18

Dudás, I (1982) Determination of contact curves and wrapping tool surface at machining of helicoid surfaces (in Hungarian), 5th Conference on Tools and Tool Materials, Miskolc, 24–7 August, GTE, II.15.1/1-15/13

Dudás, I (1983a) Verfahrensmethoden zur Berchnung und Herstellung von Hohlflankenschneckengetrieben. 6. Vortragstagung mit Internationaler Beteiligung, Fertigung und Gütesicherung im Zahnradgetriebebau, Magdeburg, T H Otto von Guericke, 28–9 September, pp 186–90

Dudás, I (1983b) Vereinfachte Herstellung und Qualitätsbeurteilung der Zylinderschneckengetriebe mit Bogenprofil Publ. TUHI, *Machinery*, **37**, pp 135–56

Dudás, I (1984) Voproszü sznabzsenija insztrumentami, geometrii proizvodsztva i takzse klasszifikacii pri proizvodsztve szovremennüh reduktorov, *Harkov Kaf. Rezanie materialov i rezsuscsie insztrumentii*, 15 October

Dudás, I (1987) Schleifen von Verzahnungen mit Superharten Werkzeugen. X. Naukowa Szkola Obrobki Sciernej, Wroclaw, 16–20 September, pp 204–7

Dudás, I (1988) Schleifen von Schraubenflächen, 7th INTERGRIND International Conference on Grinding, Materials and Processes, Budapest, 15–17 November, pp 384–95

Dudás, I (1994a) *Profiling devices of grinding wheel for geometrically correct manufacturing of helicoid surfaces*, Computational Mechanics Publications, Southampton, Boston, ISBN 1 853123676, ISBN 1 562522914, Mechatronics Conference, Budapest, 21–3 September, pp 28–34

Dudás, I (1994b) Forming of driving pair bearing patterns for worm gears, 4th International Tribology Conference, AUSTRIB '94, Perth, Australia, 5–8 December, II, pp 705–9

Dudás, I (1995) Representation of helicoid surfaces in the CAD/CAM gearing and transmissions, *The Scientific Journal of Association of Mechanical Transmission Engineers*, assisted by Gearing Committee of IFTOMM, Izhevsk, Moscow, pp 52–6

Dudás, I (1996a) Umweltfreundliche bohrtechnologien, *GEP*, Miskolc, **10**, pp 32–7

Dudás, I (1996) Generation of spiroid gearing, The 7th International Power Transmission and Gearing Conference, San Diego, California, 6–9 October, pp 805–11

Dudás, I (1999) Optimization and manufacturing of the spiroid gearing, 4th World Congress on Gearing and Power Transmission, Paris, 16–18 March, pp 377–90

Dudás, I and Bányai, K (1988) Bestimmung der Formgenauigkeit von Schneckenoberflächen mit Koordinatenmessmashine. VII. Oberflächenkolloquium mit Internationaler Beteiligung, Karl-Marx-Stadt, 8–10 February, pp 58–62

Dudás, I, Bányai, K and Bajáky, Zs (1993) Application of coordinate measuring technique for qualification of helicoid surfaces (in Hungarian), Miskolc, 31 August–1 September, pp 400–8

Dudás, I, Szabó, O and Gruzo, J (1997) Temperature variation due to the sliding of atomic planes at micro-cutting, 4th International Colloquium Mikro- und Nanotechnologie, TU Wien, 26 November, pp 41–6

Dudás, I, Tolvaj, Bné, Varga, Gy and Csermely, T (1998) Measurements applied at the experiments of environmentally clean drilling manufacturing operations, 4th International Symposium on Measurement Technology and Intelligent

Instruments, Miskolc, Lillafüred, Hungary, 2–4 September, pp 487–92

Dudás, I, Varga, Gy and Bányai, K (1996) Bearing pattern localisation of worm gearing, _VDI-Gesellschaft Entwicklung Konstruktion Vertrieb_, International Conference on Gears, Tagung, Dresden, 22–4 April, pp 427–41

Feisel, A (1985) CNC-Abrichten von schneckenprofilen, _Präzicions-Fertigungstechnik aus der Schweiz_, September, pp 91–5

Gansin, W A (1969) Slifovannije cservjaki spiroidnoj evolventnoj peredacsi, _Sztanki i Instrumenti_, No 5, pp 25–7

Gavrilenko, V A and Bezrukov, V I (1976) Geometricseszkij reszcset zubcsatüh peredács, szosztavlennüh iz evolventnokonyicseszkih kolesz, _Vesztnyik Masinosztroenyija_, No 9, pp 40–4

Georgiew, A K (1963) Elementü geometriceskoj teorii spirodnüh perdac, _Izw. Wuzow. Maschinostrojenie_, No 8

Grabchenko, A I, Verezub, N V, Dudás, I and Horváth, M (1994) High technologies in manufacture engineering, International Conference, Polytechnical University of Kharkov and University of Miskolc, Miskolc, 2 March, pp 31–6; ISBN 963 661 2382

Gyenge, Cs, Chira, A and Andreica, I (1995) Study and achievements on the worm gears, Proceedings of the International Congress – Gear Transmissions, Sofia, Bulgaria, **3**, pp 48–51

Hermann (1976) _Pfauter Werkzeugmaschinenfabrik Ludwigsburg Pfauter Wälzfräzen_, Springer-Verlag, Berlin, Heidelberg, New York

Hunt, K H (1978) _Kinematic Geometry of Mechanisms_, Clarendon Press, Oxford

Kawabe, S (1980) Generation of NC commands for sculptured surface machining from 3-coordinate measuring data, Fumihiko Kimura and Toshio Sata, Faculty of Engineering, University of Tokyo, _Annals of the CIRP_, **29** (1), pp 369–71

Kipp, G and Bielefeld (1985) Fertigung von rotoren für schraubenverdichter, _Werkstatt und Betrieb_, **118** (3), pp 157–60

Klingelnberg Firm (1972) Schneckengetriebe – Prüfgerät, Modell PSR 500 für die einzel und Sammel-Fehlerprüfungen und Schneckenrädern Werk Hücheswagen

Klocke, F and Knöppel, D (1995) Kühlschmierstoffreie Zahnradfertigung, Bericht zur 36. Arbeitstagung Zahnrad – und Getriebeuntersuchungen in Aachen

König, W and Meiboom, H M (1979) Abrichten von profilschleifscheiben für das zahnflankenschleifen, _VDI-Z_, **121** (21), pp 1087–92

Kubo, A (1978) Stress condition, vibrational exciting force and contact pattern of helical gears with manufacturing and alignment errors, *ASME J. of Mech. Design*, No 100, pp 77–84

Lierath, F and Dudás, I (1998) The modern measuring technique as the device of the effective quality assurance of the machine production, 4th International Symposium on Measurement Technology and Intelligent Instruments, Miskolc, Lillafüred, Hungary, 2–4 September, pp 465–73

Linke, H and Börner, J (1996) Precise results of stress concentrations in toothings, *VDI Berichte*, No 1230, pp 397–408

Litvin, F L (1969) Die beziehungen zwischen den krümmungen der zahno berflächen bei räumlichen verzahnungen, *Zeitschrift für Angewandte Mathematik und Mechanik*, **49**, pp 685–90

Litvin, F L (1975) Determination of envelope of characteristics of mutually enveloping surfaces (in Russian), *Mathematics*, **10** (161), pp 47–50

Litvin, F L (1989) *Theory of Gearing*, NASA RP-1212

Litvin, F L (1996a) Application of finite element analysis for determination of load share, real contact ratio, precision of motion, and stress analysis, *Journal of Mechanical Design, Transactions of the American Society of Mechanical Engineers*, **118** (4), pp 561–7

Litvin, F L (1996b) Kinematical and geometric models of gear drives, *Journal of Mechanical Design, Transactions of the American Society of Mechanical Engineers*, **118** (4), pp 544–50

Litvin, F L, Chen, N X and Chen, J-S (1995) Computerised determination of curvature relations and contact ellipse for conjugate surfaces, *Computer Methods in Applied Mechanics and Engineering*, **125**, pp 151–70

Litvin, F L and Feng, P-H (1997) Computerised design, generation, and simulation of meshing of rotors of screw compressor, *Mechanism and Machine Theory*, **32** (2), pp 137–60

Litvin, F L and Gutman, Y (1981) Methods of synthesis and analysis for hypoid gear drives of 'formate' and 'helixform', *Journal of Mechanical Design, Transactions of the American Society of Mechanical Engineers*, **103**, pp 83–113

Litvin, F L and Hsiao, C-L (1993) Computerised simulation of meshing and contact of enveloping gear tooth surfaces, *Computer Methods in Applied Mechanics and Engineering*, **102**, pp 337–66

Litvin, F L and Kim, D H (1997) Computerised design, generation and simulation of meshing of modified involute spur gear with localised bearing contact and reduced level of transmission errors, _Journal of Mechanical Design, Transactions of the American Society of Mechanical Engineers,_ **119**, pp 96–100

Litvin, F L and Kin, V (1992) Computerised simulation of meshing and bearing contact for single enveloping worm gear drives, _Journal of Mechanical Design, Transactions of the American Society of Mechanical Engineers,_ **114**, pp 313–16

Litvin, F L, Krylov, N N and Erikhov, M L (1975) Generation of tooth surfaces by two-parameter enveloping, _Mechanism and Machine Theory,_ **10** (5), pp 363–73

Litvin, F L and Lu, J (1995) Computerised design and generation of double circular arc helical gears with low transmission errors, _Computer Methods in Applied Mechanics and Engineering,_ **127** (1–4), pp 57–86

Litvin, F L and Seol, I H (1996) Computerised determination of gear tooth surface as envelope to two parameter family of surfaces, _Computer Methods in Applied Mechanics and Engineering,_ **138** (1–4), pp 213–25

Litvin, F L, Wang, A and Handschuh, R F (1996) Computerised design and analysis of face-milled, uniform tooth height spiral bevel gear drives, _Journal of Mechanical Design, Transactions of the American Society of Mechanical Engineers,_ **118** (4), pp 573–9

Minkov, K (1986) Mehano-matematicsno modelirane na hiperboloidni predavki, Disszertacija (Doktor na technicseszkie nauki), Sofia

Molnár, J (1969) Experimental and analytical method for determination of machining error occurring from the displaceability of the production system (in Hungarian), university doctoral dissertation, Miskolc, p 67

Müller, H R (1959) Zur ermittlung von hüllflächen in der räumlichen, _Kinematik Monh für Mathematik,_ **63**

Munro, R G (1991) Single flank transmission testing of screw compressor rotors, 1st International Conference on Screw Compressor Design and Manufacture, Kazan, USSR

Osanna, P H (1998) Newest development in technology, 4th International Symposium on Technology and Intelligent Instruments, Miskolc, Lillafüred, Hungary, 2–4 September

Parsaye, K and Chignell, M (1988) _Expert Systems for Experts,_ John Wiley and Sons Inc, New York

Parubec, V I (1982) Ob ekszpluatacionnüh preimusesztvah cservjacsnoj peredacsi szcilindricseszkim cservjákom, _Obrázovannüm Torom Vesztnyik Masinosztroenyija,_ No 10, pp 20–4

Patkó, Gy (1998) Dynamical results and applications in machine design, Theses of Halibitation, Miskolc, p 86

Pay, Eugen (1996) Technological experiments on the grinding process of worm gears, IMEC 96, Manufacturing Engineering: 2000 and Beyond, Connecticut, USA, August, pp 189–91

Pfeifer, T (1998) Integrated quality control, 4th International Symposium on Measurement Technology and Intelligent Instruments, Miskolc, Lillafüred, Hungary, 2–4 September, pp 365–70

Predki, W, Jarchow, F and Haag, P (1996) Self-locking worm gear units, *VDI Berichte*, No 1230, pp 129–60

Redeker, W and Radford, W F (1983) Flexible automatisierung beim profilschleifen, *Werkstatt und Betrieb*, **116** (6), pp 323–30

Rohenyi, V (1980) *Gear Drives* (in Hungarian), M Szaki Könyvkiadó, Budapest, p 628

Saari, O E (1954) Speed-Reduction Gearing, US Patent No 2,696,125

Saari, O E (1960) Skew Axis Gearing, US Patent No 2,954,704

Saari, O E (1972) Gear Tooth Form, US Patent No 3,631,736

Seol, I H and Litvin, F L (1996a) Computerized design, generation and simulation of meshing and contact of modified involute, Klingelnberg and Flender type worm gear drives, *Journal of Mechanical Design, Transactions of the American Society of Mechanical Engineers*, **118** (4), pp 551–5

Seol, I H and Litvin, F L (1996b) Computerized design, generation and simulation of meshing and contact of worm gear drives with improved geometry, *Computer Methods in Applied Mechanics and Engineering*, **138** (1–4), pp 73–103

Seireg, A A (1969) *Mechanical Systems Analysis*, International Textbook Company, Scranton, PA

Sheveleva, G I, Volkov, A E and Medvedev, V I (1995) Mathematical simulation of spiral bevel gears production and meshing processes with contact and bending stresses, Gearing and Transmissions, Russian Association of Mechanical Transmission Engineers, Moscow.

Stadtfeld, H J (1993) *Handbook of Bevel and Hypoid Gears: Calculation, Manufacturing and Optimization*, Rochester Institute of Technology, Rochester, New York

Sterki-Gunter, E (1986) CNC-gesteuertes messzentrum für kleine und mittelgrosse zahnräder, NC-Technik

Stout, K J and Blunt, L (1996) A route to standardization of 3D surface topography – characterization and functional verification, 3rd ISMTII, Japan, **10** (194)

Stout, K J, Blunt, L and Wang, K (1998) The engineered surface – relationship between metrology and the engineered surface, 6th ISMEKO Symposium, Wien, Austria, 8–10 October, pp 781–92

Su, D and Jambunatthan, K (1994) A prototype knowledge-based integrated system for power transmission design, in _Advancement of Intelligent Production,_ ed Eiji Usui, Elsevier Science, BV, pp 45–50

Szirtes, T (1988) _Applied Dimensional Analysis and Modelling,_ McGraw-Hill, New York

Terplán, Z (1965) Dimensioning problems of planetary gear drives (in Hungarian), academical doctoral dissertation, Miskolc

Weck, M (1981) Neuere steuerungskonzepte zur funktionserweiterung fertigungstechnischer einrichtungen, _Industrie Anzeiger,_ **12**, pp 11–20

Weck, M, Escher, Ch and Beulker, K (1996) Simulation and optimization of generation of tooth flank modifications, International Conference on Gears, Dresden, Germany, 22–4 April, pp 883–95

Zak, P Sz and Sapiro, I I (1984) Modelirujusije masinü dlja cservácsnüh peredacs, _Vesztnyik Masinosztroenyija,_ No 12, pp 6–8

Zhang, Y, Litvin, F L and Handschuh, R F (1995) Computerized design of low-noise face-milled spiral bevel gears, _Mechanism and Machine Theory,_ **30** (8), pp 1171–8

Zotow, B D (1961) Osi zaceplenija spirodnüh peredac, _Izw. Wuz. Masinostroenije,_ No 6

INDEX

A

Active surface 3, 16, 17, 75, 81, 107
Actual profile 201, 203, 206, 250
Archimedian helicoid surface 149, 163, 213
Axial backing-off 108
Axial clearance 262, 266
Axial pitch 62, 144, 172, 173, 200, 206, 217, 221
Axial section of the wheel 59, 142
Axial section 4, 6, 15, 16, 18, 23, 33, 34, 35, 36, 37, 41, 42, 59, 61, 62, 72, 87, ,97, 100, 119, 122, 127, 136, 142, 143, 144, 146, 147, 162, 170, 171, 175, 181, 183, 187, 201, 204, 206, 217, 221, 238, 239, 246, 278, 281, 291
Axial thread parameter 105, 106, 107, 108, 163

B

Back surface 103, 105, 106, 107, 108, 109, 116, 118, 119, 148, 162, 164, 165, 166, 167, 169, 173, 257, 289
Base profile 115

C

CAD 3, 71, 148, 228, 251, 290
CAM 3, 71, 290
CAQ 3, 290
Characteristics 14, 15, 27, 28, 33, 44, 59, 72, 76, 85, 87, 103, 142, 143, 200, 201, 204, 206, 209, 214, 217, 227, 228, 230, 237, 240, 260, 266, 271, 272, 279, 291, 292

CIM 3, 223, 229, 290
Circular profile worm gear drive 21, 25
CNC grinding machine 257
CNC wheel-truing device 250
Coefficient of profile displacement 232
Conical Archimedian helicoid surfaces 149
Conical helicoid surface 26, 75, 76, 87, 96, 97, 100, 104, 105, 108, 109, 124, 125, 127, 136, 148, 165, 169, 246
Conical worms 3, 26, 29, 61, 102, 124, 224, 255
Conjugated surfaces 15, 239, 289
Conjugated worm 114
Constant lead cylindrical helicoid surface 115
Contact area 18, 40, 41, 170, 228, 230, 232, 233, 238, 240, 242, 260, 261, 262, 263, 264, 265, 266, 267, 285
Contact axis 24
Contact curve 16, 17, 23, 51, 58, 59, 66, 67, 68, 69, 70, 85, 92, 94, 101, 126, 127, 142, 143, 145, 146, 147, 148, 149, 155, 180, 210, 230, 231, 232, 233, 239, 292
Contact curves and relative velocity 230
Contact point 17, 52, 58, 142, 209, 210, 211, 212, 213, 217, 239
Contact surface 14, 16, 25, 81, 95, 114, 191, 227, 228
Continuous wheel control 257

Convolute helicoid surface 79, 155, 156

Coordinate transformation 36, 37, 44, 209

Cylindrical helicoid surface 16, 30, 35, 61, 64, 65, 115, 124, 127, 162, 163, 246

Cylindrical worm gear drives 26, 29, 33, 140

D

Diagonal relief 106, 107, 169

Direct generation 77

Direct method 290

Drive housing 238, 284

E

Edge ribbon 113

Edges 42, 95, 103, 105, 106, 107, 114, 223

Efficiency 19, 20, 192, 223, 230, 233, 234, 271, 272, 274, 277, 279, 280, 291

Expert system 222, 223, 224, 225, 228

F

Face gear 75, 76, 77, 124

Face surface 103, 105, 114, 115, 119, 262, 269

Finite Elements Method 226

Flyblade with helicoid face surface 114

Flyblade with thread face surface 120

Flyblade 114, 120

G

Gearing ratio 27, 28, 64, 95, 98, 187, 191, 203, 226, 232, 241, 280

Generating milling cutter 30, 35, 72, 115, 162, 164, 223, 257, 286

Generator curve 64, 65, 66, 67, 105, 127, 164, 201, 204, 290

Generator of the profile 37

Geometry of generating tool 109

H

Head clearance 107

Heating curve 277

Homogeneous coordinates 37, 44, 46, 66

I

Inclination of shape 82

Indirect method 290

Intelligent automation 227, 228, 259

Intelligent Integrated System 222, 223, 224

Investigation by simulation 237, 241

Involute helicoid surface 78

K

Kinematic method 37, 292

Kinematical and controlling 258

Kinematic gearing ratio 224

Klingelnberg made 201

L

Lead 5, 19, 20, 35, 36, 42, 43, 61, 62, 64, 65, 82, 87, 103, 107, 115, 116, 117, 136, 144, 147, 172, 173, 200, 212, 213, 214, 232, 238, 241, 255, 257, 258, 262, 290

Lead angle 19, 20, 35, 42, 43, 62, 64, 82, 107, 116, 144, 147, 172, 173, 257

Lead parameter 64, 117, 212

M

Machining geometry 15, 30, 76, 109, 136, 148, 169, 246, 253, 292, 293

Manufacturing errors 238, 239

Mathematical-kinematical model 61, 64, 135, 136

Measuring machine 204, 206, 207, 209, 245, 253, 292

Meshing condition 15, 200, 260, 293

Movement parameter 58, 85, 92, 94, 96, 126, 142, 162, 230, 250

N

Noise level 227, 230, 274, 288

Normal plane 110, 111

Normal transversal 21, 24

Normal vector 52, 57, 69, 70, 92, 141, 166, 209, 210, 212, 213, 214

O

Oil film   16, 17, 233
Oil level   274, 275
Operating on mechanical principle 186
Optimal tool profile   99, 151, 154, 157, 160
Optimal wheel generation   148

P

Parametric equation   66, 103, 111, 163, 164
Pitch   1, 9, 19, 21, 31, 33, 62, 82, 103, 110, 144, 172, 173, 200, 206, 208, 217, 221, 238
Position vector   46, 51, 57, 66, 67, 71, 85
Principle of backward generation 181
Principle of enveloping   182, 183, 192
Profile angle at reference cylinder   41
Profile distortion   170, 184, 192, 246, 250
Profile in axial section   6, 33, 34, 35, 41, 63, 97, 100, 136, 146, 147, 162, 163, 171, 175, 176, 239, 278, 281
Profile of the worm   42, 181, 201
Profiling of the grinding wheel   23

R

Radial back surface   133, 164
Radial relief   107
Radial thread parameter   103, 105, 106, 107, 124
Radius of curvature   147
Range of gearing ratio   28
Relative motion   44, 51, 136, 247
Relative velocity   16, 17, 51, 52, 70, 86, 87, 92, 141, 165, 230, 231, 236
Relief   105, 109, 118, 119, 169, 172, 173, 178
Resharpening   105, 107, 113, 115, 116, 123, 170, 173, 176
Ruled surface   4, 12, 13, 15, 16, 18, 26, 28, 30, 76, 77, 79, 182, 183, 277
Running in of the drive   271
Running in   111, 234, 265, 266, 267, 268, 269, 271, 277, 280, 287, 291

S

Side edge   103, 105
Side surface   3, 38, 40, 44, 103, 113, 115, 116, 118, 119, 136, 161, 257
Spiroid drive   27, 28, 29, 76, 148, 226, 236, 244, 292
Substituting worm   114, 115
Superhard grinding wheel   30, 75, 246
Surface roughness   20, 21, 97, 187, 239

T

Theoretical profile   75, 201, 246
Thread curve fitted on reference cylinder   213
Thread grinding   102, 194, 245, 251, 255, 257, 258, 259
Thread motion   27, 51, 164
Thread parameter   103, 105, 106, 107, 108, 117, 124, 163
Thread surface   4, 42, 51, 65, 66, 67, 95, 101, 102, 116, 135, 250, 255, 256, 257, 258, 259
Tool displacement   124
Tool profile   35, 41, 72, 87, 94, 95, 96, 97, 99, 122, 135, 136, 151, 157, 160, 169, 192, 250
Tool surface   3, 65, 66, 67, 68, 69, 70, 71, 95, 109, 127, 132, 255, 290
Total length of the contact curves 234
Transformation matrix   39, 46, 51, 64, 66, 67, 105
Truing grinding stones   3
Type of drive   9, 12, 18, 25, 233

U

Undercutting   26, 42
Universal thread grinding machine 255, 258, 259

V

Vector function   66
Velocity vector   17, 51, 52, 87, 92, 165, 231, 236

W

Wear of wheel   246, 250

Wheel profile   42, 43, 44, 75, 76, 97,
    101, 102, 136, 143, 145, 147, 148,
    151, 160, 161, 169, 171, 182, 184,
    188, 189, 217, 223, 227, 246, 250,
    252, 254, 258, 289

Wheel truing   250, 251, 258, 259,
    289, 290, 292

Wrapping curve   14, 44

Wrapping surface   13, 14, 15, 44, 59,
    68, 77, 107, 108, 163

Printed and bound by CPI Group (UK) Ltd, Croydon, CR0 4YY

08/10/2024

01042187-0002